YOUR FARM
IN THE
CITY

YOUR FARM IN THE CITY

LISA TAYLOR
AND THE GARDENERS OF seattle tilth

An urban dweller's guide
to growing food
and raising livestock

BLACK DOG
& LEVENTHAL
PUBLISHERS
NEW YORK

Published by
Black Dog & Leventhal Publishers, Inc.
151 West 19th Street
New York, NY 10011

Distributed by
Workman Publishing Company
225 Varick Street
New York, NY 10014

Manufactured in Canada

Cover and interior design by Red Herring Design

ISBN-13: 978-57912-862-3

h g f e d c b a

Library of Congress Cataloging-in-Publication Data available upon request

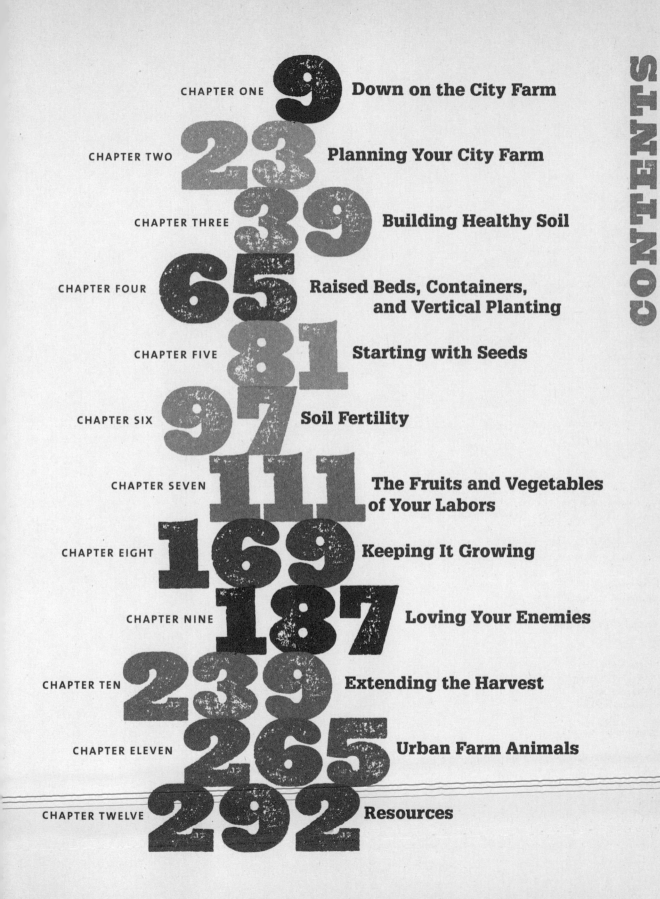

CONTENTS

DOWN ON THE CITY FARM

**Y JOURNEY AS A CITY** farmer began with a visit to a garden unlike any I'd ever seen. A riot of color and activity, it was filled with bugs, creative trellises, and cool garden art. Fruits and vegetables were tucked among perennial herbs and shrubs, rather than in long, straight rows. Not only was it peaceful and inspirational, this garden was educational! Everywhere I looked were signs and demonstrations about how I could grow food at home and help the environment by gardening organically.

Located on a 12-acre city park, tucked behind formidable holly hedges, the original Seattle Tilth learning garden at the Good Shepherd Center in Seattle is an edible oasis in a frantic, bustling city. What a refreshing change to the monotony of lawns that predominate in urban and suburban landscapes. One visit and I was convinced: I wanted to create a mini-Tilth garden at my home. I envisioned a yard filled with vegetables and herbs, with fruit trees along the fence line and grapevines climbing an arbor over our porch swing. I wanted a place where I could pick my dinner just moments before it came into the kitchen. And that's exactly what I have now.

You can create a city farm, too! Described in this book are the simple, straightforward, and effective techniques that we have refined and taught at the Seattle Tilth learning garden for more than thirty years. Although my experience growing food and raising chickens has been in Seattle, the techniques described in this book can be used to create a high-yield city farm across the United States and Canada.

We'll help you learn how to grow your own food and keep small livestock, such as chickens, goats, or bees. You'll explore how to grow food in a variety of urban settings, such as backyards, patios, balconies, and community gardens. With a little reading and a lot of effort, you'll turn your boring, nonproductive space into a beautiful, diverse, food-producing city farm. So let's get started.

What Is Urban Farming?

Urban farming is a new term for growing a kitchen garden (or, to use the French term, *potager*) which has been practiced since the advent of cities. Previously, these efforts were called everything from pea-patching to victory gardening to edible landscaping. For our purposes, *urban farming* is defined as resourceful food gardening in a city setting and *farmers* are those nobly growing food for themselves, their families, and their community. Unlike the back-to-the-land movement of the 1970s, this new wave of urban farming isn't about giving up the city life and heading back to the farm but about embracing the notion of the little family farm in the big city.

The recent surge of interest in eating locally has inspired city dwellers to see food growing potential all around them. A grassy lawn with a southern exposure is a wasted opportunity for self-reliance

Good Food NOW!

We can thank the good food crusaders for this renewed interest in urban food production—Alice Waters, Michael Pollan, Jamie Oliver, Frances Moore Lappé, Barbara Kingsolver, and the slow- and local-food advocates. These modern-day good food evangelists honor the work of the organic movement founders, those smart, brave folks who always questioned the industrialization and mechanization of agriculture. Pioneers, such as Rudolf Steiner, Lady Eve Balfour, Sir Albert Howard, Masanobu Fukuoka, Alan Chadwick, and J. I. Rodale, recognized almost immediately how inappropriate the industrial, human-made model was when applied to a natural system. Documentaries, including *Food Inc.*, Jamie Oliver's *Food Revolution*, and Eric Schlosser's *Fast Food Nation* are opening more eyes to the perils of the disconnect between people and their food. Even Oprah Winfrey has featured good food advocates several times on her show. For the first time since Eleanor Roosevelt, there is a vegetable garden in the White House lawn!

and independence. Those front yards are perfect farmsteads. In fact, any patch of ground (whether it is covered by grass or pavement) that gets six to eight hours of sunlight and has access to water is a potential urban farm. In formerly vacant lots, it is now common to hear the exuberant clucking of laying hens or to smell the sharp tang of raised-bed herb gardens.

So whether you have a small patio or a quarter-acre, you can grow good food close to home; practice organic farming techniques that improve the environment; and make productive urban lands that currently provide little benefit for people. Plus, if you are brave and municipal zoning laws allow it, you can expand your farm to include livestock, such as chickens, bees, goats, ducks, and rabbits.

For many people, the goal is to grow healthy food, close to those who will eat it, and to have control over how it is grown. Others want their

kids to forge a connection with nature and to know that their food doesn't come from boxes on the supermarket shelf. For those with more space, city farming makes produce available to communities that have no source of fresh food— especially inner city "food deserts." Farming in the city also fulfills the basic need for people to connect with growing things and the life force of our planet.

The aim of this book is to teach you how to grow food, but it is also about eating. Linked inextricably with city farms is the culture of preparing and preserving fresh food. By farming, you are becoming part of your local food system, following your food from seed to table to compost pile. So as you plan your farm, start first with what you love to eat. For me, there is nothing finer than lounging in the backyard, listening to songbirds, and reaching up to pick a few raspberries right off the vine and pop them "al fresco" into my mouth!

Why Grow Your Own Organic Food?

The organic urban farm: More than a source of sustenance

People grow their own food for different reasons—to save money, to avoid chemicals, and for excellent taste, variety, and freshness. But the organic urban farm does so much more than provide a source of dinner.

Environmental concerns

Organic methods improve the urban environment by building healthy soil, encouraging biodiversity, and creating a habitat for wildlife. Healthy soil also serves as to filter to clean storm water runoff. By growing food right outside our front doorstep, we also reduce our use of gas and our carbon footprint.

Economic issues

Growing your own food organically can save time and money. Hours at the store and in the car are

How Much Can You SAVE?

So how much money can you save by growing vegetables on your city farm? Claims of reaping substantial savings can be a bit like weight-loss disclaimers: Actual results may vary. What you can expect to save depends on what you grow, the size of your garden, your skill as a grower, and the weather.

Still, growing your own food does make economic sense. For example, in a tiny 1.5-square-foot garden bed, you could grow six nice-sized heads of lettuce. If you spend $3 on a six-pack of lettuce starts and another buck on fertilizer and water, you could expect to harvest at least six heads of lettuce. If you grow a loose-leaf variety and pick the outer leaves as they mature, the plant will regrow for several harvests before it finally bolts to make seed. Fancy, organic heirloom lettuce (if you can find it at the market) goes for $3 a head—so for as little as $4 invested, you could produce $18 worth of greens. And you won't have to make six trips to the market.

Grow a bigger garden and you'll save even more. Michelle Obama's 1,100-square-foot White House kitchen garden was started for $200 and produced 1,000 pounds of produce—enough to feed folks at the White House, with extra going to food banks. Roger Doiron of Kitchen Gardeners International, and his wife Jacqueline tracked their production from a 1,600-square-foot garden and estimated that for an initial $282 investment, they grew about $2,390 worth of fruits and vegetables.

replaced by time invested in your farm. A modest urban garden can be created in a couple of weekends and, once established, can be maintained in a few hours each week. Growing organically saves money spent on expensive chemical fertilizers and pesticides.

Food gardeners support local businesses, such as nurseries, seed houses, and tool suppliers, and they contribute to municipal compost programs. Buying locally supports the local economy and keeps money in the community. It is estimated that only 15 cents of every dollar that is spent at those big national box stores stays in the local economy, while more than 45 cents of every dollar spent at a local business stays in the community—this is economic soil building.

Food safety, food security

In this era of food recalls, homegrown produce reduces the dangers of giant industrial agriculture. Recently, tomatoes were pulled out of markets for fear of contamination, which turned out to be caused by jalapéno peppers. I followed this on the news as I ate a fresh spinach salad with tomatoes and peppers from my own garden.

Family and community fun

Tending a garden teaches children stewardship and a deep appreciation of the wonder, beauty, and magnificence of the Earth. Urban farming is also a great way to expose our children and our community to the natural world. Building an urban farm teaches kids responsibility, increases self-esteem, and gives them a sense of purpose. Plus, kids love to dig, plant seeds, and search for creatures.

Besides, growing food is fun, and eating what you've grown is a hands-on culinary adventure. Kids will eat almost anything they can pick from the garden. For 15 years, I have led children and their parents on edible sensory walks in the Seattle Tilth Children's Garden. Although my powers of persuasion should not to be overstated, parents are often amazed at how many different, seemingly non-foodlike plants their kids will eat and enjoy when they are invited to participate in the growing process. Even the pickiest eaters will find themselves taking a nibble of an edible flower or crunching a handful of fennel that they picked fresh from the garden.

Nature-deficit disorder

Working together on your urban farm can help to reverse the effects of nature-deficit disorder. This term, coined by author Richard Louv, refers to the fact that most children (and adults) in the U.S. spend little time outside. Louv claims that modern mass media—with its 24-hour news cycle of murders, assaults, and abductions—has created a culture of fear. Parents are scared to leave their children alone outdoors. So we keep them indoors, playing Farm Town video games instead of exploring the natural world.

Your city farm is a safe place for unstructured exploration and a great way to connect with life cycles and the seasons. It teaches children where their food comes from and how it is grown. It also promotes the development of imagination and cultivates a love of the natural world.

Feelings of independence and self-sufficiency

Growing food is a life skill that teaches planning, careful observation, and patience. It is also profoundly liberating! With every harvest comes a great feeling of independence and self-sufficiency. City farmers may never need to survive solely on the food they can grow, but it is important to know that they can directly provide for themselves and their families.

Creating a multipurpose, productive environment

A city farm is an edible, interactive landscape. Replacing landscape bimbos—plants that look good but produce little else—with ones that provide food, a habitat for creatures, color, scent, texture, and beauty creates an environment that serves more than one purpose.

Exercise

Tending a city farm is physical work in fresh air and natural light, and comes without the cost of exercise equipment. Money saved on a gym membership can be used for good tools and soil-building amendments.

Also, part of the joy and satisfaction of growing and eating homegrown food is the work that goes into it. After laboring through the season on a vegetable garden, we really appreciate those tomatoes at the end of the season. We find ourselves eating with reverence, slowing down to savor them and celebrating a successful harvest.

Although urban farming is work, it needn't kill you. In later chapters we'll show you how to work wisely so that more can be accomplished with less wear and tear on your body.

Enhancing your emotional and mental health

The health benefits of gardening extend beyond eating more fresh fruits and vegetables. For many city farmers, gardening is a therapeutic way to relax, stay calm, and better cope with the demands of a busy urban lifestyle.

Working in the garden relieves stress and boosts your mood. Surrounding yourself with soothing colors, smells, and the sounds of nature has an uplifting effect on your mind and spirit. Recent studies show that exposure to soil bacteria increases serotonin levels, the neurotransmitter that produces a feeling of well-being. Growing food for your family also gives you a sense of accomplishment and increases your feelings of self-worth.

Reducing your carbon footprint by keeping it local

To grow, process, ship, and store food from across the world consumes nonrenewable resources and exacerbates global climate change. In fact, much of what we eat has traveled an average of 1,500 miles! Urban farms lessen our reliance on cars by replacing trips to the store with footprints to the backyard.

Growing for flavor

Most vegetables and fruit sold in supermarkets have been designed and chosen for shelf life rather than flavor. Conventionally grown produce is picked early and "ripens" on the way to the store. Many of the most delicious varieties don't keep or ship well. Since city farmers plant what they love, selection can focus on flavor and variety. Produce picked when it is ripe tastes much better than food ripened en route. Suddenly, you'll find yourself eating a lot of healthy food that hadn't appealed to you before.

Convinced? Let's Get Started!

A case for taking it slow

In the excitement of starting a city farm, many inexperienced growers jump in with both feet. Huge parcels are optimistically planted by folks with great intentions but little experience in what it takes to manage a food-producing garden. I have

THE SELF-TAUGHT GARDENER

It is natural to want to know everything about gardening from the start, but this is impossible. The field of horticulture is so vast that you will never know everything. You could spend your whole life learning and only grasp a fraction of everything there is to know. Focus on what really interests you. Find the plants or areas of agriculture that you are drawn to and learn all you can about them; this will keep you busy. Take classes, read books, and try things in your garden to build your knowledge and know-how.

KEEPING A GARDEN JOURNAL is

a great way to learn about gardening, but be sure to buy a few good books for your reference library and consult local resources, such as your county extension service.

Include books about aspects of gardening you want to learn more about—irrigation, permaculture, edible landscaping, medicinal herbs, plus field guides to wild plants, weeds, and insects. Not all books work for all gardeners. To find the right books for you, check them out from your local public library before making a purchase. (See the resources chapter for a list of books that I find especially useful.)

heard people say that last year was their first time growing a garden and this year they have tripled their garden space and aren't going to buy groceries. While noble and inspiring, this passion and enthusiasm is a slippery slope. There is an old saying that you should never plant more than your partner can weed and water.

Weather and other demands of urban life can sidetrack even the most thoughtfully planned garden. Spending loads of time and money to experience nothing but failure is disheartening and stops many people from trying again the following year. If you're relatively inexperienced, you should start a city farm incrementally, with the primary goal being to learn the art and craft of agriculture (as well as getting a few good tomatoes).

The Tilth Way—Where We're Coming From

The "Tilth Way" takes many techniques used by the different schools of sustainable agriculture and combines them in a way that's helpful for the small-space gardener, the backyard farmer, and the urban grower. The techniques described here

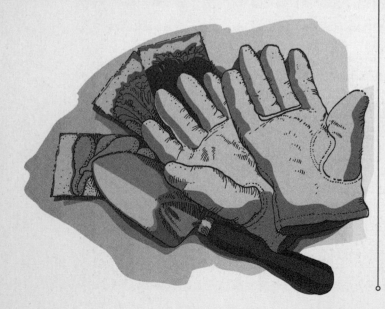

are not new: They are a combination of traditional farming methods and recent innovations adapted to small-space urban growing. We emphasize techniques for year-round food growing. Ours is an experiential learning model that acknowledges that most people learn best by doing and more deeply by teaching others.

Inherent in our methods are the tenets of organic gardening: building healthy soil, working with nature, encouraging biodiversity, and conserving resources by being thrifty and resourceful to produce the most from a small space. We use intensive gardening practices to get the highest yield from small spaces; these include double digging, raised beds, companion planting, vertical growing, intensive planting, and season extension. For established beds, we advocate a low-till method that lightly tills soil before planting, so that roots have space to grow without disrupting microbial life or soil structure.

We are always striving toward a "closed gardening system," which includes the following key elements:

- *recycling organic matter as compost*
- *using local and organic materials*
- *incorporating recycled materials as functional or artistic elements in the garden*
- *growing perennial plants along with annual vegetables to support beneficial insects and wildlife*
- *seed saving*
- *minimizing outside inputs with the goal of creating a diverse and sustainable urban ecosystem*

Organic Gardening Tenets

Build healthy soil

Healthy, living soil is the foundation of any productive city farm. Although soil may seem like the least interesting thing about growing plants, this is where it all starts. Improving your soil is the single most important thing you can do. Underlying all organic gardening is the recognition that the health of the soil and the health of all living things are inextricably linked.

Good soil grows healthy plants that resist disease and pests. Healthy soils absorb and retain more moisture, requiring less water. In chapter 3, we'll show you soil-building techniques, including composting, adding organic matter as mulch, and growing cover crops. Then in chapter 6, you'll also learn about soil fertility and how using organic, slow-release fertilizers supports the millions of microorganisms that play an essential role in healthy plant growth. Healthy soil is the key building block of any successful garden.

Work with nature

Rather than seeking to tame or control nature, the organic farmer works in concert with natural cycles. You are helping the garden grow, not making it grow. Try to strike a balance between leaving nature alone and making the changes needed to grow crops. This takes careful observation and an

understanding of the natural forces that are at work around you.

Rather than fret about the aphids covering your roses, try to see how these soft-bodied creatures fit into the larger ecosystem. Aphids are unsightly and sticky, but they seldom kill plants. Rather, they are essential food for predators and beneficial insects—they are the plankton of the garden. Without aphids, there would be no food for ladybugs, birds, or other helpful insects. In chapter 9, you'll learn to distinguish the good bugs from the bad and how to help control them naturally.

Right plant, right place

Plants will thrive with minimal care when they have the ideal growing conditions. Pests and diseases are opportunistic; that is, they are always looking for an easy mark—typically a weak plant growing in the wrong place. A shade-loving plant that needs wet feet will not thrive in full sun and sandy soil.

Likewise, a tomato plant that needs heat, sun, and well-drained soil may grow in a wet, shady spot but will not produce fruit. In chapter 2, you'll assess the characteristics of your property to determine what plants will thrive on your site, in your soil type, and in your microclimate.

Encourage biodiversity

Nature is incredibly diverse. Natural areas contain hundreds of different living things that create a balance between those who eat and those who are eaten. Imitate these natural areas by including many different types of edible plants in your city farm.

Biodiversity also means also including annual and perennial plants that provide different heights, textures, and colors of flowers and foliage. This results in a beautiful and interesting landscape that better mimics the canopy layers in natural areas. Birds and bugs inhabit and use different canopy

Forgotten Pollinators

You don't need a butterfly bush to attract pollinators to your farm. Anything that flies or crawls spreads pollen. Bees and butterflies are the darlings of the pollinator group, but there are countless winged creatures that move pollen from plant to plant. Let some annual vegetables go to seed and while they bloom, check out all the critters that find their way to the flowers for their color, scent, and sweet nectar.

Flowers are specially adapted to attract specific pollinators in order to make seeds and further their species. Pear blossoms are beautiful signals of spring, but they stink like rotting fish or dog poop. This stinky beauty isn't a ploy to keep things away; it is beckoning the pollinators, like filth flies, that are attracted to rotting things.

Include a flower border along your vegetable garden to provide food and habitat for the myriad pollinators that help make it all possible. I sometimes let my collards make flowers and watch the flies, bees, and hummingbirds sip. Then I eat the blossoms in a salad. They are somehow more delicious having been danced on by a butterfly.

DON'T PANIC!
BE FLEXIBLE
MONITOR & ADJUST
HAVE FUN

layers—some use the understory, some the upper canopy, while others move among the different layers. Including plants of different heights creates these layers and encourages more diverse bird and bug populations, leading to a healthier, more sustainable urban farm ecosystem.

Use the least toxic approach and see what happens

This is the simplest and yet the most challenging tenet of organic gardening. The least toxic approach to any problem in your garden is to do nothing and see what happens. Sometimes it is hard to repress the knee-jerk reaction that bugs on your plant are a bad thing. We have been conditioned to fear or hate anything that crawls. Even though I know that

over 95 percent of the insects in the garden are beneficial and won't hurt my plants, I still have a subconscious urge to squish any bug I see. Nothing can be done about creatures or plant problems until you positively identify the cause of the problem. Then weigh the creature's damage to the garden versus its benefits and decide what to do.

In chapter 9, you'll learn about the life cycle of creatures to better understand how they fit into the garden ecosystem.

Conserve resources: Water wisely

Using water wisely conserves this most important and precious resource so that it can be used by people, plants, and wildlife. Too much or too little

SEATTLE TILTH

SEATTLE TILTH—Learn. Grow. Eat. These three words capture the essence of this hands-on, educational organization. Seattle Tilth has led the way in teaching folks about organic and sustainable ways to grow food since 1978. Our mission—to educate and inspire people to garden organically, conserve natural resources, and support local food systems in order to cultivate a healthy urban environment and community—comes to life in numerous learning gardens scattered throughout the greater Seattle metropolitan area.

Class participants and garden volunteers learn about city farming topics that range from starting their first garden or compost pile to keeping chickens, bee or goats in the city to food preservation and canning techniques for a bountiful harvest. Even indoor workshops are up close and personal, where teachers with experience and knowledge to share might bring a chicken or miniature goat to the classroom!

Seattle Tilth's children's garden programs started in 1988, and since then have helped more than 50,000 children and 30,000 adults catch the bug, so to speak, of growing food. Kids learn about insects and how they help the garden, as well as take part in digging, planting and harvesting. School classrooms visit the garden as part of their science curriculum and summer campers keep the gardens growing year-round.

Other areas of focus include providing economic development through the creation of green jobs and market opportunities, organizing community events, mobilizing a stellar volunteer corps and participating in coalitions working to make good food available to all members of the community. Gardening for life is the great equalizer.

LEARN MORE:
seattletilth.org.

LEARN. GROW. EAT.

seattle tilth

water can weaken plants, making them more susceptible to pests and disease. Give plants the right amount of water and make every drop count. In chapter 8, I'll show you how to use drip irrigation so that you water the plant (not the sidewalk) and to water your garden early in the morning or late in the evening to reduce evaporation from wind and sun.

Learn as you grow

There is no better way to learn how to grow plants than to do it: Reading gardening books, even really good gardening books, is no substitute for putting theories or principles into practice. Keep a journal to track and test your observations. Pay close attention. Use your senses and your observational skills to their fullest to learn what is going on in your environment—quite frequently the answer to your question is right in front of you.

Whether you intend to farm to supplement your kitchen pantry or you hope to create something bigger that might provide food for your community, this book will give you practical ideas and time-tested techniques to create and maintain a high-yield organic city farm. Throughout are techniques for maximizing space and production while conserving resources and protecting the environment. Included are profiles of growers doing incredible things and ways children can help on the city farm. Hopefully this will encourage you to start growing your own little farm in the big city.

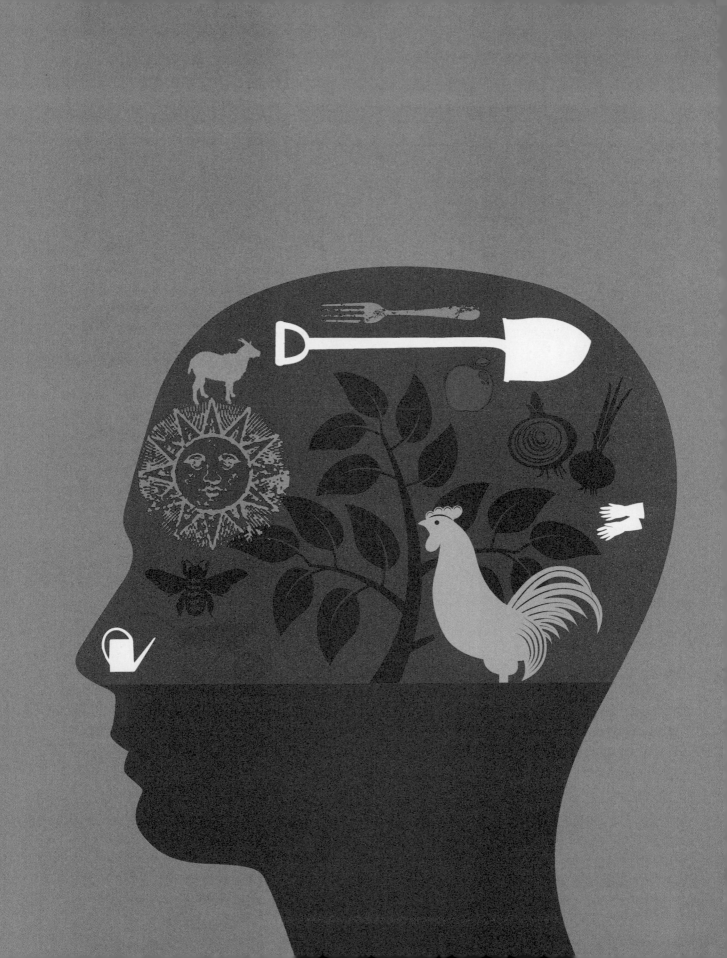

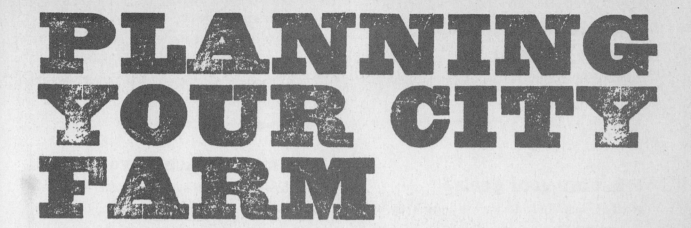

PLANNING YOUR CITY FARM

Before you start planning, it pays to spend some quality time dreaming. This is one of the most exciting parts of starting your city farm. What will your ideal farm look like?

 URVEY YOUR TERRAIN: EACH sunny spot is a potential place to grow food. Imagine your front yard with herbs and blueberries, grapevines growing over the carport, beds bursting with salad greens, a sunflower fence along a boundary line, and an herb spiral filled with vegetables. Maybe your backyard has a vegetable garden and a chicken coop, with a beehive tucked back in a corner. In place of the current blocks of cement or patchy grass is an abundant, productive landscape that provides good food and healthy activities for you and your family. At the end of each season, your cupboards are filled with beautifully canned food and your freezer loaded with ready-to-eat veggies from the garden. This, my friends, is the dream of city farming.

These dreams lead to the practicality of a plan. What will you include on your farmstead? Will you grow food in raised beds, containers, or more traditionally in rows? Is a small orchard in your future? Will you integrate edibles into your landscape, mixing vegetables and edible flowers among your perennial beds? Will you raise chickens or keep bees? Dream big, with no holds barred.

Creating Your City Farm

You've thought about what would be an ideal farm for you. Now it's time to determine what's a realistic farm for you and your lifestyle.

What are your goals?

What do you want to accomplish on your city farm? What do you want to grow? Is this a start-up year, when you learn a little about growing plants and maybe eat a tomato or two? Or are you more experienced and want to grow enough to feed the family? Being clear about what you want from your city farm can make it easier to stay focused and achieve your goal—to create an edible urban oasis. My general goal is always the same: "I want to eat from our yard year-round." However, before each planting season, I generate a set of smaller goals. These are usually plant- or site-specific. This year I'm focusing on my lettuce, which is always bitter. I'm going to learn more about lettuce so I don't keep growing plants that we can't eat! This helps me to improve each season.

How much space do you have?

Vegetables need a lot of sun. Look around for spots that are level, get six to eight hours of sunlight, and are easy to access. If possible, wait until landscape trees and shrubs have leafed out to make sure a spot is sunny enough. These are likely places for your garden. Do you have room on a deck or balcony to grow herbs or vegetables in containers? Is the best place for your garden actually in the front yard? These are all broad considerations before you begin.

Maybe you want a more landscaped front yard. You can include edibles, such as fruit-bearing trees,

RETHINK YOUR LAWN

There is nothing finer than lazing about on a sunny afternoon on the lawn, digging your toes into the soft grass. But often a lawn in a city is difficult and expensive to maintain. Imagine instead replacing those sunny patches of non-productive, resource-sucking lawn with a lush ecology teeming with life-producing food for your family and community.

Granted, some lawn is desirable and needed. It is hard to have a leisurely picnic or game of badminton in a vegetable garden. But while planning, consider ripping up some of your lawn and growing something edible. Under that non-productive carpet of grass is soil waiting to be liberated!

shrubs, and medicinal herbs along your paths and around a little seating area. You have to determine what works for you.

Is it legal?

Before you make big plans for chickens and a parking strip packed with produce, check your municipal land use and traffic codes. It would be a shame to get the family excited about, say, raising chickens, only to find out later that there are restrictions against keeping livestock in your municipality. There may also be building codes about how close to the property line you can put toolsheds or compost bins. Having to move an established chicken coop or potting shed would certainly be a drag.

If you live in a housing development with a covenant or a homeowners' agreement, check about keeping animals

and where a vegetable garden is permitted. Although some homeowners' association agreements prohibit residents from putting a vegetable garden in the front yard, most would not have a problem with a tidy herb garden in containers. If you're the subversive type, consider mixing some medicinal plants, edible flowers, and fruit trees or shrubs (like plums, apples, or blueberries) within traditional landscaping. That way you will have an edible, productive landscape that won't get you thrown out of the neighborhood.

How much time do you want to spend gardening?

I love to garden and spend time outside working in my yard, but I also live in a city that is bursting with great things to do. Personally, I don't want so large a garden that it feels like a chore. A small garden may need serious attention a few times a week or 20 to 30 minutes of care each day. Consider how much time you want to spend in the garden to determine its size. Remember: Chickens, goats, and rabbits also need daily care, just like having a pet.

How much can you eat?

Even the most committed city farmer may find it challenging to keep up with the harvest after the garden starts booming. Plant a mix of vegetables that can be eaten fresh with those that preserve well. Sprinkle in some that can be harvested over time. During the summer, we don't eat much fresh kale because we have plenty of squash, cucumbers, beans, peas, and lettuces to fill our plates. We blanch and freeze the kale we harvest throughout the season and feast on these hardy greens from the freezer all winter in soups and as side dishes.

How much money do you have to spend on your garden?

Decide before you begin how much you can afford to spend on your operation for the season. If you already have some tools, a garden hose, and decent soil, a small vegetable garden can be planted for less than $30. However, installing a complete landscape design or building a fancy chicken coop can be quite expensive.

If one of the reasons you are growing food is to save money, be realistic about how much you can save. Figure out how much you currently spend on food that could instead be grown in your own garden. This is a good way to begin a budget. Now also consider long-term investments, such as tools, pots, hoses, and the like. Write down a dollar amount and stick to it. It is very easy to get into the thrill of gardening and buy lots of tools, plant starts, birdfeeders, garden art, and gadgets, and go way over budget!

What will you grow?

Make a list of food you like to eat—this will determine what you want to grow. Don't make the list too large. Growing several things well is more satisfying than trying to grow everything with minimal success. In later chapters, we'll show you what does well in your region and space restrictions, as well as plants that are heavy producers or can be harvested over time.

Will you include animals?

Are chickens or ducks in your future? Do you have enough space for a couple of miniature dairy goats? Do you plan to set up a beehive to produce homegrown honey? If you would like to keep small animals on your farm, think about where they will live in your yard and how much space they will need. Chapter 11 will help you learn about keeping small livestock on your city farm before you make the plunge!

Will you preserve the harvest?

Thrifty city farmers freeze, can, and dry their garden bounty so that they eat vegetables from the garden all year. (See chapter 10 for information about preserving your harvest.) Others share garden produce with friends and coworkers. Befriending a neighbor with gifts of fresh eggs could make up for that clucking hen!

Designing Your Farmstead

Now that you have imagined what your city farm will be like, it's time to make your plan. This section will help you assess your site and design your garden.

Start a garden journal

When you begin to design your city farm and vegetable garden, it is also a great time to start a garden journal. This is an important tool for planning, making observations, and recording crop rotation. This is not English Composition class. The journal is for you alone and can be whatever you need it to be. I was initially reluctant to keep one. However, since I started keeping track, my yields are bigger and my timing for planting is better. My garden now has more variety and produces more food over a longer period. So I can't argue with success.

Your journal doesn't have to be fancy—I use an inexpensive spiral notebook because it is portable and the pages don't fly around the garden when there is a gust of wind. Yours might include maps of your planting areas, details about when and where seeds are sown, and plant growth rates or how much was harvested. You might use your journal frequently, making daily entries and detailed notes about weather, plant health, and compost pile ingredients. Or you might make just a few notes to record a growing season. The goal of keeping a journal is to improve your farming skills every year and to learn from your experience.

Possible Things to Put in Your Garden Journal

Sketches of your garden beds

What you planted

Crop rotation maps

Sowing and planting dates

Sun/shade patterns

Landscape zones

Your observations about what went well

Theories about why something succeeded or failed

Planting calendars

Harvest dates and how much was harvested

Seed lists

Compost and other soil amendments you have used

Observations about weather

Frequently used local resources, such as your county extension service

Articles from newspapers and magazines

Ideas for next season

section, you'll go through the process of assessing your site.

Some people learn about their space by observing and digging around in it. Another option is to create a site map, where you plot the different aspects of your yard. Both methods work and will give you a good idea about what you are working with.

To begin, get to know the plants that are already in your yard. They aren't going anywhere unless you dig them up and move them. Rather than removing them, figure out what edibles you can add to your existing landscape. Perennial beds or mulched areas under trees can become places to grow vegetables or edible flowers and shrubs.

Don't be afraid to rip plants out or move them to better suit your needs. Your farm does not have to be hamstrung by the previous tenant's landscape choices. But be sure to really think out your plan before you undertake any major plant moves or removal. It is easy to imagine moving plants around like furniture, but it is quite another thing to actually unearth them and plant them somewhere else. It takes skill to cut down large trees and moving or digging up larger perennials may be difficult, even gargantuan in scope! Besides, ugly hedges might be

Your Urban Ecology

Getting to know your place

Before you can really make a successful garden plan, you have to get to know your space. By understanding your yard, you'll put plants where they will thrive and can be cared for easily. In this there for a reason; those arborvitae may be blocking severe winds or the view of the hideous apartment building next door!

Inherited landscapes

A perfect city farm would start with a clean slate, but urban sites are inherited landscapes. You adopt what the previous tenant has left behind. Your inheritance may include large ornamentals or pernicious weeds. The previous owner may have left a toxic legacy of chemical fertilizers or a mature asparagus patch. You'll never know what you have until you look around. If you're not sure what's a weed and what's a plant, ask a neighbor with a garden—gardeners love to talk about weeds. If the soil in an area smells funny or nothing seems to grow there, be careful. This may be a place where a car has been parked. We'll talk in chapter 6 about soil tests and how to make sure your soil is safe.

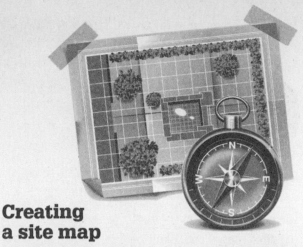

Creating a site map

You are going to need a map, whether it is a rough sketch or graphed to scale. You may end up creating a couple of different maps, depending on the project.

Since our garden beds are scattered about our yard, I arranged them on my map so I could see them all at once. There are four beds in the front yard, a bed along the side of the house, and six beds and various containers in the back. I sketch all my planting areas on one page in my garden journal and then write what and when I plant different things. This helps me keep track of crop rotations and planting times.

Taking the rhodie for a walk

When we first started to create our city farm, there was a large rhododendron blocking the pathway around one side of our house. We had heard that rhodies can tolerate being moved and that they love nothing better than a walk around the house. This seemed simple enough.

When we started to excavate around the rhododendron to dig it up we realized this wouldn't be as simple as rearranging the furniture. The soil was compacted and hard, requiring pick axes and lots of muscle. Because of low hanging branches, much of the work of digging this thing up was spent on our knees and bellies. When we did finally manage to dislodge the roots, we were faced with the daunting task of moving this 350-pound monster across the yard to its new home. Using a push-and-pull method assisted by tarps and a hand truck, we finally managed to move it the 100 feet only to be faced with digging another gigantic hole so it could be transplanted!

Needless to say, now we are less quick to think about ripping stuff out. Instead we think about how we can make what we have work for our city farm.

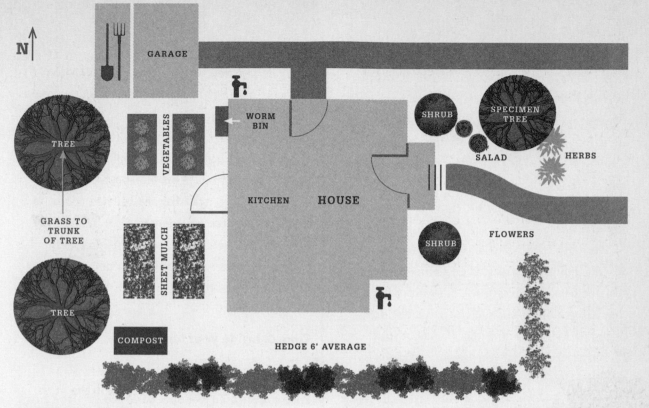

To sketch your site

- Mark the cardinal directions (north, south, east, west).

- Sketch the shape of your property.

- Draw the approximate size and location of your house or other buildings.

- Include sidewalks, patios, and driveways.

- Draw existing landscape elements, such as trees, shrubs, and garden beds, and how much space they take.

- Map driveways, alleys, or other routes for getting materials onto your site.

- Include roadways and sidewalks that border your property.

- Locate water sources, gutter downspouts, and electrical outlets.

To draw a map to scale

- Use graph paper to plot permanent features of your landscape. Your scale may be one square equals one square foot or a smaller scale if you have a large area to map.

- Mark the cardinal directions, date, and scale. For example, ¼ inch = 1 foot.

- Measure the area of your site with a long tape measure.

- Measure the area of buildings, patios, decks, sidewalks, and driveways.

- Measure the distances between the house and the property line.

- Measure the area between decks, sidewalks and driveways.

- Draw trees, shrubs, and garden beds, and their canopy space.

- Map driveways, alleys, or other routes for getting materials onto your site.

- Specify the location of water sources, gutter spouts, and electrical outlets

- Indicate roadways and sidewalks that border your property.

- Note any building codes that have an impact on your garden.

Location, location, location

After you have gotten the lay of the land, it is time to place plants where they will work with your life. How do you currently use your space? What are the areas you use most? Are there places in your yard where you seldom go? Understanding how you use your space will help to identify where to put things that need daily attention and those that require minimal care.

Permaculture designers use a zone system to describe how areas are used on a property. Zones describe how we typically use our yards, what areas we use more often, and which areas are seldom visited. The idea is simply this: If you put your garden in an area that you use often, you will be more likely to care for it or harvest produce when it is ripe.

To identify your landscape zones, think about how people and animals use your space. Make note of what areas are frequently used and which areas are seldom visited. Identify the areas you live in and your movement patterns. Successful gardens are located in a convenient, frequently visited place with easy access. Knowing your traffic patterns and how you use your property will help you find the best places for high-maintenance plants, tool sheds, fruit trees, and animal dwellings. If you are creating a vegetable garden or trying to find the best place for a chicken coop, locating zones 1 and 2 is important. These are the two outdoor zones that are most frequently used.

Zone 0: Inside your house

Zone 1: Most visited

Zone 1 is the area just outside your door—the areas you can get to without putting on your shoes. Close and convenient, Zone 1 is just the right location to

ZONE	FREQUENCY OF USE	PLANTS TO INCLUDE	OTHER ELEMENTS
0	in your house	aloe, herbs	Canning and food preservation equipment
1	most visited outdoor space	kitchen and herb garden, greenhouse or potting bench	bird feeders, worm bin
2	semi-intensive	outer boundary of vegetable garden, fruiting shrubs, berry vines, crops that don't need continual harvesting	chickens, rabbits, tool shed, compost bins, wheelbarrow parking
3	large farming zone	fruit trees, large garden area for row crops, berry vines, flower borders	goats, ducks, bees, wood pile
4	minimal care	native plants, perennial cover crops, flower and wildlife borders	
5	seldom used	wild area, untended plants	area behind the shed, outer corners

dash out to snip a few herbs to throw in whatever you are fixing for dinner.

In urban landscapes this zone may extend out 15 to 20 feet. This is the area for high-maintenance plants, the best place for the kitchen garden, and a great place for a greenhouse or plant-propagation area. This is also an ideal location for birdfeeders and outdoor living areas, such as decks and patios.

Zone 2: Semi-intensive

Zone 2 is just outside your outdoor living space. You will need to slip into your garden clogs to reach this zone. Things in zone 2 are visited daily, but not many times a day (the way those in zone 1 are).

As the outer boundary of the garden, this is where you locate the toolshed, wheelbarrow parking, fruiting shrubs, vines, and crops that don't require continual harvesting, plus chicken coops or rabbit hutches.

Zone 3: Large farming zone

Getting out to zone 3 may feel more like a job, since you have to plan to go out there to do something. Found here are fruit trees and native plants. In this large garden area are row crops, berry vines, wildlife borders,

goats, ducks, and bees. These are areas that you visit for specific jobs, like harvesting fruit or working in a large vegetable garden.

Zone 4: Minimal care

Many urban sites are small and do not have zone 4 areas. Included here are things that require minimal care, such as native plants, perennial cover crops like alfalfa, flower, and wildlife borders. These are places only visited a few times each year, to apply mulch or do some weeding.

Zone 5: Areas seldom or never used

This is the area behind the shed, the wild areas, or outer corners of your property.

Zones are not laid out in radiating lines like waves. Some zones may be isolated bubbles. Location isn't all that goes into identifying a landscape zone. Sometimes the area is close to your front door but it is around the corner of the house, requiring you to slog through the mud or past a red twig dogwood that slaps you in the face. This area may be zone 5, since it is never visited because passage is difficult. I know, because I'm the one who gets slapped in the face

Can zones change

?

You bet, If you add or change something in your landscape it may change how that space functions. At my house, we seldom used the front stoop because it wasn't a great place to sit. We have an herb spiral in our front yard that is about 12 feet from the front door. It sat squarely in zone 2 and we had to make an effort to water and care for plants that were growing there. After we added an 8-foot deck out from the stoop, we use this area all the time. The spiral has moved closer to zone 1 and is easier to water and harvest.

every time I try to get back to the woodpile!

If zones are confusing or don't work for you, try to locate your garden close to your kitchen door or the main pathway through the yard. There is a direct correlation between the distance from the kitchen to the vegetable garden and your tendency to care for it. For every foot you move your vegetable garden away from the kitchen, the less likely it is that you'll make the trek to tend it. Upon later reflection, many of my gardening failures were caused by poor placement.

Access

How will you bring materials onto your site? Think about routes for moving a wheelbarrow of mulch or compost around your yard. A steep slope or steps between your pile of mulch and your new garden area could make your work more difficult. Also, think about places you would dump that load of woodchips so that it can sit there until you are ready to distribute it among your garden beds. Access to compost bins and tool storage should also be clear and easy.

Mapping the Elements

Mapping your sun, soil, water, and air doesn't have to be a grueling assignment. You will most likely become familiar with your site as you work in your garden and yard. To really get to know your space, it would be ideal to live there for a year, updating your site map with things you discover about sun, water, soil, weeds, birds, bugs, and plants. If you can't wait a season to plant, pay attention, notice as much as you can, and make notes about what you learn.

Determine your microclimates

Get to know the different microclimates in your yard. A microclimate is the climate near the ground;

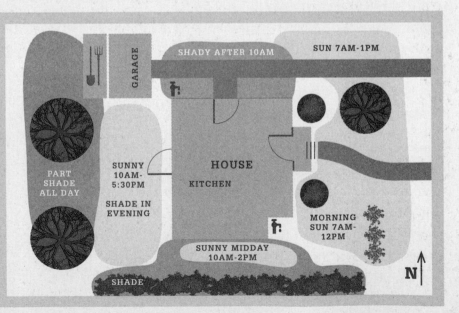

HOW TO MAP SUN AND SHADE

Pick a sunny day in May or August.

Draw a line on your site map where the shade falls and note the time.

Repeat this hourly throughout the day.

Sun patterns change throughout the year so repeat this map several times per year to get the most accurate picture.

GARAGE

SHADY AFTER 10AM

SUN 7AM-1PM

PART SHADE ALL DAY

SUNNY 10AM-5:30PM

SHADE IN EVENING

HOUSE

KITCHEN

MORNING SUN 7AM-12PM

SUNNY MIDDAY 10AM-2PM

SHADE

N

What the *heck* is a microclimate?

A microclimate is a small area that experiences its own weather conditions. Around your property there are tiny climates that are each affected by sun and wind exposure, soil type, and houses, fences, and other landscape features. Soil types influence how quickly a garden bed will warm up. Sandy soil warms up more quickly than heavy, silty soil, which also takes longer to dry out. Likewise, raised beds will warm up more quickly than beds that are sunken. It may take you a few seasons to fully understand and identify your unique microclimates.

To pinpoint different microclimates, walk around your yard on a nice evening wearing short sleeves and use your skin to feel differences in heat and cold. Stand still and notice the wind and where the yard is warmer. Use visual clues to find tropical or chilly areas. Look for frost pockets—areas where frost or dew lingers longest, or other areas that seem to heat up more quickly. Snowmelt can help you visually see and identify colder microclimates—in sheltered cold spots, snow may take several weeks longer to melt than in more exposed, warmer areas.

Observing where early weeds first appear can help you identify the locations that warm up more quickly; these may be ideal places to plant early spring crops.

it is affected by air, soil, water, sun, and other landscape features. Climates changes slightly around your site. These slight changes can affect how well crops grow. A front yard could have half a dozen different microclimates. Finding warmer or cooler pockets in your yard will help you know where to put heat-loving plants and those that can tolerate more cold. Microclimates can change with the addition or removal of landscape elements.

Sun

If you want to make sure an area gets the six to eight hours of required sunlight for growing vegetables, you can map the sun/shade patterns for that area.

Soil

As you work on your city farm, you will become familiar with the different types of soil in your yard. Urban soils are notoriously compacted, which affects how well water will drain. (For more information about soil, see chapter 3.) Make general observations about your soil. Notice how water reacts on your soil's surface and then dig down to see what is happening deeper. Identify areas that hold water and stay boggy, places that are sandy and dry out quickly, and spots where nothing seems to grow. Get a spade and dig around in different areas where you're considering locating your garden. Notice how your soil changes from place to place. Look at the plants that are doing well on your site—especially weeds, since they will grow wherever conditions are favorable. You can make notes about your soil on your map. This will be a handy reference when you are deciding where to put plants.

Water

Mark water sources on your site map. Placing your garden beds close to a water source will make your job easier. A few days of hand watering may be fine when seedlings are starting, but that kind of commitment will be difficult to maintain throughout the season. Make sure your hoses are in good repair so that water is always available where you need it. Save water by using sprinklers with a pattern that matches the shape of your beds or drip irrigation.

Air

All plants need some wind, which helps them to develop strong roots and stems. However, most vegetable crops and fruit cannot tolerate strong wind. Pay attention to any areas that seem particularly windy or are more sheltered. Notice the direction in which the wind blows most of the time.

If you notice strong winds, consider including some kind of windbreak in your plan. Fences with open slats and hedges are the most effective windbreaks, since they absorb or slow wind down while allowing some air to pass through. Solid walls or fences create an eddy of intensified wind on the calm side, which can damage plants more than if there were no windbreak at all.

Pollution

If you live on a busy street, auto exhaust and other airborne pollutants are valid concerns. Invisible particulates can land on the surface of your vegetables, making them mildly toxic. The areas near busy streets and around the foundation of older homes may have high lead levels from particulates and paint chips that collect in the soil. If possible, locate your vegetable beds away from sidewalks, busy streets, and the foundations of buildings.

RAPID RETURN TO FREE WIND SPEED

TURBULENCE

GRADUAL RETURN TO FREE WIND SPEED

Keep the lead out of your garden and home.

Here are some ways to reduce exposure to lead and airborne pollution:

- *Place garden 50 feet away from heavily traveled streets and at least 10 feet from painted structures.*

- *Take routine soil tests for lead.*

- *Grow in raised beds or containers filled with clean soil if lead levels are higher than 1,000 parts per million.*

- *Remove and dispose of the top two inches of soil with high lead levels — lead is relatively immobile and stays in the top couple inches of undisturbed soils.*

- *Use border plants and hedgerows along the perimeter to absorb particulates.*

- *Grow leafy greens and other hard-to-wash vegetables in areas away from the street.*

- *Wash vegetables thoroughly with a couple drops of dish soap or in a vinegar/water solution (2 ounces vinegar to 1 gallon of water).*

- *Lay down mulch to keep particulates from reaching the soil.*

- *Wash your hands to avoid secondary contamination from touching and ingesting contaminated soils.*

- *Keep the dust down by removing shoes when indoors and using rugs to catch soil that is tracked in from outside.*

- *Eat a healthy diet which is high in calcium and iron but low in fat to reduce absorption of lead into the body.*

Creating Your Design or Layout

Making a plan

You have decided what you want on your city farm and you have the lay of the land. Now it's time to make a plan. Plans help us get things done. Your plan needn't be formal: Think of it as a road map that will give you some direction as you head into your city farming adventure. You may have more than one plan—you may have a long-term plan for converting your yard into an urban oasis and a seasonal plan about what you will grow this year. Look to it when you lack inspiration or wonder what to do next.

Your plan should work for you and contain as many details as you need to stay on track. It may include site maps, a budget, plant and seed lists, information about resources and supplies, sketches, cuttings from magazines, bed maps, planting times, and long-term goals/desires. Or it may be as simple as a rough sketch and a plant list.

Overwhelmed by All the Possibilities?

When you're first getting started, there are so many things to do that it can feel overwhelming. It takes time to establish a vibrant urban landscape, but you will be rewarded for your efforts with fresh produce and an abundant, productive environment. Think of this as a project that will come to fruition over time. Identify the most important things to do this year. Then carefully add new elements so that everything fits together to create something bigger—a real city farm. Have fun and learn as much as you can. As you expand your operation, plant only as much as you can weed, water, and eat.

DON'T PANIC!

ASK YOURSELF:

What do I want most?
What will it take to do that?
What can I manage right now?

What to start with:

- *Plant a small vegetable garden and a few herbs*
- *Start composting*
- *Harvest and eat what you grow*
- *Learn as much as you can*

Lay out your garden

Setting up an inviting garden environment is as important as where you put the beds. Start by thinking about where plants will grow and places where people will go. Regardless of whether you will garden in raised beds or containers, you have to be able to easily reach your garden. If you can't reach the plants, it will be difficult to harvest them. If getting water to your garden regularly is awkward, it will be hard for seeds to germinate. Lay out your garden so that paths are clear and beds are accessible.

I like a spacious garden that invites me to walk in and interact with the plants. By spacious I don't mean a large garden, just one that is open and inviting, and easy to move around in, even if it is physically small. Paths should be uncluttered and

easy to see. Plants should be easy to reach without straining. If you have to tiptoe down a tangle of tiny paths to get to your vegetables, you will be less likely to spend time there. If beds are so wide that you have to do contortions to reach in for harvesting, a lot of what you grow will never make it into the kitchen.

To design an inviting and accessible garden, keep beds narrow so that plants can be reached from both sides. Paths should be easy to navigate. Create beds that are 24 to 30 inches wide with paths on both sides and rows no longer than 8 or 10 feet long. Long rows make for more walking, which can feel like extra work. It is also difficult to reach plants in beds wider than 30 inches. If you can only reach the bed from one side, narrow it down to 18 inches or the distance you can comfortably reach across without leaning onto the soil. If you already have a garden bed that is wider than 3 feet, consider ways you can add stepping-stones to make reaching your plants easier.

Paths should be 18 to 24 inches wide so that you can work and move without crushing plants. If children will be working in your garden make the beds narrower (12 to 18 inches) and the paths wider (24 to 30 inches). Wider paths also give you space to kneel and sit while working.

Most of us water our gardens with a hose (until we finally break down and put in a drip irrigation system!), so think about how you will get your hose down your paths to your plants. Hose guides can help keep plants at the ends of your beds from being crushed and whipped. A sturdy pole or piece of rebar pounded into the end of your bed will keep plants safe. I use large planting containers at the end of my beds; they make excellent hose guides and increase my growing space.

Sample farmsteads can be found in the resources chapter.

No yard? No worries

Don't let the lack of yard and soil stop you from growing your own food. Creative city farmers grow vegetables, herbs, and even small fruit trees in containers and boxes on rooftops, balconies, asphalt parking lots, and patios. (See chapter 4 for creative ways to grow.) If you want to garden and connect with your neighbors, rent a plot in a community garden or pea patch. With regulations loosening about keeping livestock in urban areas, many community gardens are morphing into farms, complete with chickens and rabbits, which provide community members with eggs, manure for fertilizer, and endless hours of entertainment.

That's a lot to think about. Getting to know your site and making your plan is a process. Don't be scared by all the details. Learning about the ecology in your yard is fun! This is the beginning of your journey as a city farmer.

PATHS FOR KIDS and FAMILIES

If you are designing your garden beds with children in mind, you will need to reverse your thinking about space. Invite kids into the garden by making the space suited to their unique characteristics and needs. Children take up a lot of space as they learn to move and control their bodies. Paths should be two feet wide and covered with a unique identifiable material — straw or burlap sacks work well (wood chips are too pointy and sharp) so it is easy to see where plants and people go. Garden beds should be narrow (12 to 18 inches wide)—the perfect width to jump over. Kids will always take the most direct route in the garden and this doesn't always mean using the paths. They like to jump over the beds. If your family garden beds are 2 feet wide or wider, you will be constantly fighting the kids to "stop jumping in the beds." Create a garden that fits your family rather than trying to change the kids.

Profile:

LEARN MORE:
urbanfarming.
org.

What if healthy food could come from places like the vacant lot down the street, the local school, or the lawn in front of city hall? That's the goal of Urban Farming, one example of the many nonprofit organizations that help address the issues of hunger and food security.

Since it started in 2005, when the organization created three gardens in Detroit, Urban Farming has helped to orchestrate gardens in many other cities and on many different kinds of sites, including container gardens at shelters in Los Angeles and a school garden in Chinatown, New York.

Urban Farming knows that gardening is a multifaceted task: It not only provides food for the table, but it also teaches skills, including math, project planning, construction, and even small-business administration. The Urban Farming Entrepreneurship and Money Management Program helps young people and adults learn how to grow not just vegetables, but also a business. Vineyards Growing Veggies encourages homeowners to include food crops within ornamental gardens, and the community gardens set up through Urban Farming's program provide free food for the neighbors.

If the typical 20-by-20-foot Urban Farming plot seems like too much space to handle, participants can take advantage of the Urban Farming Food Chain, a vertical gardening project. The walls of buildings, such as the Los Angeles Regional Food Bank, have sprouted with panels growing tomatoes and other edibles; the harvest is donated to those in need.

BUILDING HEALTHY SOIL

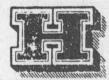ERE AT SEATTLE TILTH, we know there is a big difference between dirt and soil. Dirt is what you wash off your car or your hands. Soil is alive. It is filled with living things and supports other living things, such as plants, animals, and people.

Soil is a living thing, so we need to care for it as we care for all living things. I know I wouldn't thrive if I were walked on or didn't get anything to eat. One of the first lessons we teach in the Seattle Tilth Children's Garden is that, in a garden, people and plants don't belong in the same place. When we step on the soil, we compact it, squishing out space for air, water, roots, and creatures. So watch where you step.

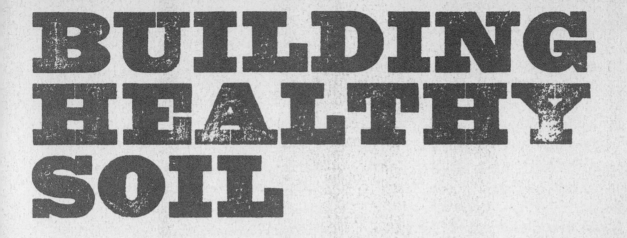

Children in the Garden: Where Are Your Feet?

If your children will be helping cultivate and care for your city farm, take time to teach them how to be in the garden. A garden is different from being at the park or playing in the yard. In a garden, people and plants don't go in the same place. Plants live in the soil. People can walk on the grass, sidewalk, stepping-stones, and garden paths. Remind each other as often as needed to "watch your feet" or ask, "Where are your feet?"

Your Soil

What is soil? Think of it as a pie. One-half of your soil is space for air and water. Almost all of the other half (45 percent) is inorganic matter—sand, silt, clay, and rocks. Just a little slice (2–5 percent) of your soil is organic matter. This is a small piece that makes the difference between dirt and soil. It is also the glue that holds it all together and the engine that makes plants grow.

This small slice of the pie is filled with living things, including plants, animals, and microorganisms. In a teaspoon of good garden soil, there are more than 3 billion microorganisms. The smell of the forest floor is really the micro-organisms—dirt alone doesn't have a smell. These tiny creatures make the difference between dirt and soil. Micro-organisms make minerals and nutrients in the soil and air available to plants. In return, plants give the microbes the energy they need to live. Without these microbes, your plants can't utilize the nutrients in your soil.

Many current landscaping practices overlook feeding the living part of the soil. In the woods, plants don't need to be fertilized or mulched. Nature is messy. Plants are designed to break apart, constantly dropping pieces of organic matter, which decompose and feed other plants. At some point we started to apply our housekeeping aesthetic to our landscapes. We are always raking up and removing leaves and other plant debris. Though I like a tidy house, I don't expect that level of cleanliness from my landscape.

Plant debris is food for the creatures in the soil. Decomposers, such as

Organic Matter 2-5%

Inorganic Matter 45%

Space 50%

Soil is living and needs the same things as other living things, including the following:

SUN. There are places in your yard that get more sun and other places that are shady.

AIR. Depending on your soil type, there might be big air spaces or tiny air spaces. You can increase these spaces by digging to loosen the soil.

WATER. The water in your soil will change depending on the season and your soil type.

FOOD. What does your soil eat? Typically, other living things, like leaves, twigs, and dead insects.

MICROBIAL MAGIC AT THE Root Zone

Plants can only anchor themselves in dirt. They can't make use of any mineral nutrients without the help of some really cool fungi. There is an area around the root zone where mycorrhizal fungi gather and bring minerals from the soil in a form that plants can use. The plants return the favor by giving the fungi carbohydrates. This is an amazing symbiosis that benefits both the plant and the fungi. Since these fungi help plants absorb mineral nutrients from the soil, many gardeners inoculate their beds with mycorrhizal fungi before they plant. You can get dried mycorrhizal fungi through sources online. (See Resources in chapter 12.)

worms and sow bugs, change leftover organic material into rich humus that feeds the plants. These bits also create a carpet that suffocates weeds, eliminating competition for water, soil, and nutrients. This mulch layer helps to hold water, protects roots, and provides habitat for the microorganisms that help plants grow. It's beautiful and simple. When we try to keep our landscapes unnaturally tidy, we rob our soil of the food that makes for a healthy, vibrant ecosystem. Without organic matter, we just have dirt.

Nonliving Soil Elements

Later in the chapter we'll cover all you need to know when using organic matter to improve your soil and garden. Before we do that, let's look at the part of your soil that isn't alive. The nonliving part of your soil is composed of a combination of sand, silt, and clay. Soil is classified simply by particle size. Understanding your soil type will let you know what you're working with and how to make it better.

The best way to get to know your soil is to put your hand in there and feel the grit. The largest particles are sand and the smallest are clay, with silt falling in the middle. Compost particles are large, they are very porous and shaped irregularly, with lots of convex areas to hold water and air. Soil particles are very small, but the difference in particle size between the different soil types is really big. If sand were a city bus, then clay would be a grain of rice, and silt would be a beach ball. Compost particles would be the size of a midsized car. The size and shape of soil particles affects how pore space is arranged and how water reacts to your soil. Let me describe the different types of soil.

Soil texture

SAND

Sand particles are gigantic. They are smooth and round, like miniature boulders. Sand feels coarse and gritty. Grains of sand are between 0.06 and 2 millimeters in diameter. They don't fit together tightly. There is a lot of space for air and water. Sandy soils are easy to dig. It is easy for water to percolate through sand, but it dries out quickly. Sandy soils are low in mineral nutrients.

SILT

Silt particles are much smaller than sand. They range in diameter from 0.0039 to 0.0625 millimeters. Silt particles are small enough to be carried by water and deposited on the sides of creeks and rivers. Silt particles are irregular in shape and fit together more tightly than sand. Silty soils often hold water and are spongy. They feel gluey when wet. Silt sticks together when wet, but falls apart easily when poked. Silt particles sometimes look like black pepper has been sprinkled in the soil. Even though it can be really spongy, when silt dries out it is still easy to dig. Silty soils have moderate nutrient levels.

CLAY

Clay particles are tiny—so tiny you can't even feel them. Pure clay feels like powder. Clay particles are less than 0.0039 millimeters in diameter, less than a fraction of the width of a human hair. Clay particles are all shaped the same, so they fit together tightly, like bricks. The space between these tiny particles is very small. Clay is sticky, soft, and pliable when wet, and hard as a rock when dry. Since it takes time for water to percolate down into the tiny pore spaces, clay soils often shed water. When clay does get wet, it takes a long time to dry out. It is hard and impenetrable, like pottery when it dries. Clay soils are very high in nutrients, but are often so hard that plants can't access them through the tightly packed soil.

You probably don't have pure sand, silt, or clay. You have loam, which is a mix of all three, with sand or clay being predominant. If you want a visual picture of the amounts of sandy, silt, and clay take a sediment test (see below).

Homegrown soil tests

These tests for soil texture are easy to do and will help you get to know your soil better. There are two different ways to test your soil texture—the squish test and the sediment test. These can be performed safely by children with no special equipment. If you are looking for a science fair project, your child could test and record the soil texture in your back and front yards. They could demonstrate how to test for soil moisture the day of the fair. (See chapter 4 for moisture testing.) Your child's teacher would love it!

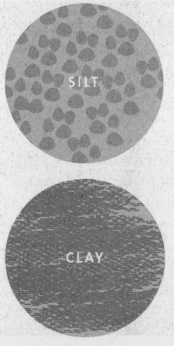

SEDIMENT TEST

If you are curious about how much sand, silt, or clay you have in your soil, take a sediment test and see how the three are distributed.

Materials

Wide-mouth mason jar
Some garden soil
Water

SQUISH TEST

Materials

A shallow pan or dish tub

Some garden soil

Water

Instructions

1. Collect a cup or two of soil by taking 6- to 8-inch-deep slices of soil from the area you want to test.

2. Remove really big rocks.

3. Mix in just enough water so that it starts to stick together. If the soil is dripping, it is too wet. In that case, just add more soil until it makes a moist dough that just sticks together.

4. Now take a small handful of moist soil and try to form a worm in your hand.

5. Work the soil dough between your thumb and fingers to form a cylinder.

Sandy soils might hold the shape for a minute but will break apart easily.

Silty soils will form a short worm, but will break apart when poked.

Clay soil will form a worm over 2 inches long. If you can rub your soil between your hands and make a longer worm, you have a lot of clay in your soil.

Instructions

1. Put one cup of soil in a wide-mouth mason jar.

2. Fill the jar with water, leaving about an inch of space at the top.

3. Put on the lid and ring.

4. Shake so that the soil is dispersed in the water (a few seconds).

5. Set the jar on a flat surface. Watch the particles settle. Sand will collect first, then silt. Clay will be suspended in the water, and organic matter will float on top.

6. Let the water settle overnight or longer. There should be a very fine layer of clay on the surface. The water may still be cloudy. It may take several days for these tiny clay particles to settle.

Measure the different layers and guesstimate the percentages of each to identify whether you have sandy or clay loam.

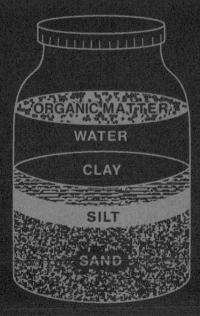

ORGANIC MATTER

WATER

CLAY

SILT

SAND

WEEDS AND WHAT THEY SAY ABOUT YOUR SOIL

Weeds can be an indicator of what is going on in your soil. Weeds are advantageous and will grow wherever the conditions are right. Knowing the soil conditions that weeds favor will help you to identify your soil type.

Bindweed, field
(*Convolvulus arvensis*)

hardpan surface; light sandy and acidic soil

Chickweed
(*Stellaria media*)

cultivated soil with high fertility or low fertility if plant is yellow and lacks vigor

Creeping buttercup
(*Ranunculus repens*)

wet, poorly drained clay or silty soils

Dandelion
(*Taraxacum officinale*)

heavy clay and compacted soil; acidic soil; common in lawns and cultivated soils

Dock
(*Rumex*)

poorly drained, waterlogged, acidic soil

Horsetail
(*Equisetum arvense*)

light, sandy, acidic soil

Lamb's quarters
(*Chenopodium album*)

cultivated soil with high fertility or low fertility if plant is yellow and lacks vigor

Moss
(*Selaginella*)

waterlogged, poorly drained, acidic soil

Pigweed (*Amaranthus retroflexus*)

cultivated soil with high fertility or low fertility if plant is yellow and lacks vigor

Plantain (*Plantago*)

heavy clay, poorly drained, acidic soil; common in lawns and cultivated soil

Why urban soils are so awful

Many new urban farmers complain that they have really poor soil. This is often because city soils do not occur naturally. Here's what happens when a site is developed for residential housing. First, all the topsoil is scraped off. Then heavy machinery drives all over the site and workers stomp around on the soil for several months. When the house or apartment is finished, the site needs to look nicer so it can be sold. At that point, the developer smoothes out the ground, covers it with rolls of sod, plants a few landscape trees and shrubs (around here it would be rhododendron and azalea), and the dwelling is put on the market for sale.

The rolls of sod were grown on a sod farm in clay so the soil will stick to the grass roots when it is rolled up. So they lay clay on top of compacted soil. They sell it to you and you are supposed to take care of it. Developers rarely amend the soil with compost when planting either sod or other plants. It is a wonder that these plants grow at all! The good news is that you can fix this mess.

HOW MUCH COMPOST DO YOU NEED?

1 CUBIC FOOT COVERS

24 square feet
1/2 inch deep

12 square feet
1 inch deep

6 square feet
2 inches deep

4 square feet
3 inches deep

1 CUBIC YARD COVERS

648 square feet
1/2 inch deep

324 square feet
1 inch deep

162 square feet
2 inches deep

108 square feet
3 inches deep

Compost

Building healthy soil is surprisingly easy. If you want to grow strong plants that taste great and resist pests and diseases, add compost. How do you improve heavy clay soil? Add compost. Want to make your soil retain water better, so plants don't dry out? Add compost. Compost really is a miracle worker. It boosts the health of soil, helps plants combat disease and pests, and helps conserve water.

In this section you'll learn how to use the organic matter on your site to improve your soil. You'll also learn different techniques for composting yard and food waste, mulching, and growing your own fertilizer using cover crops

Using compost to improve the soil

As I've mentioned, the best way to improve soil is to add compost. If you don't have time to make your own, you can buy commercial compost in bags or have it delivered in larger quantities to use in your garden. This is a great way to start building healthy soil.

Add compost

Measure the area you'll be working with and determine how deep you will spread the compost. Then do the math to figure out how much compost you'll need (see table). Since compost does not have high levels of nitrogen, it will not "burn" your plants, so you can plant immediately after you add it to your garden beds.

- *For new garden beds, mix 2 to 4 inches of compost into the top 8 to 12 inches of soil.*

- *To improve established garden beds, dig 1 to 2 inches of compost into the top 8 to 12 inches of soil.*

- *Topdress with 1 to 2 inches of compost around your perennial shrubs and in existing beds where it is impossible to mix in the compost. Topdressing means to spread a thin layer of compost on top of the soil. As you water your garden, the nutrients in the compost filter down, improving your soil. Topdressing is easy and looks great.*

Backyard composting

Composting is nature's way of returning nutrients to the soil to help the next generation of plants grow. Many people gather up their yard waste, grass clippings, and fall leaves and put them at the curbside for yard waste recycling. This is great because it keeps a valuable resource out of the landfill. Many cities recycle yard waste into compost that can be used to improve soils. If you pay for yard waste to be taken away and then buy compost each spring for your garden, you are paying for it twice. Think about the organic material on your site as a resource that will help improve your landscape.

ORGANIC MATERIALS FOR YARD WASTE COMPOSTING

DO PUT IN YOUR COMPOST PILE	DON'T PUT IN YOUR COMPOST PILE
Old plants	Plants that are diseased or infested with insects
Flowers	Pernicious weeds that spread by roots or runners
Annual weed leaves, stems, and flowers	Weed seeds
Grass clippings	Meat or dairy products
Potting soil	Food waste
Livestock manure—chicken, rabbit, goat, cow, or horse waste	Poop—dog, cat, rodent, exotic bird waste
Cardboard, paper napkins, shredded paper	Evergreen leaves or needles
Sawdust or wood shavings	Branches bigger around than your thumb and longer than your hand
Twigs and sticks	Plants with thorns
Deciduous leaves	Plants sprayed with herbicides or pesticides
Straw	Sod

All organic matter—anything that is alive, anything that was alive, any part of a living thing and anything made from a living thing—will break down and become part of the soil. But not all things break down at the same rate. Focus on organic materials that will break down most efficiently in a backyard compost pile. Living in cities close to our neighbors, we want our compost to break down quickly so as not to attract rodents or create a foul smell.

One general rule: If you don't want it in your garden, don't put it in your compost pile.

YARD WASTE CHART

GREENS

Food scraps	15:1
Grass clippings	20:1
Rotted manure	25:1

IDEAL MIXTURE	**30:1**

BROWNS

Brown leaves	40–80:1
Corn stalks	60:1
Straw	80:1
Paper	170:1
Woodchips	500:1
Sawdust	500:1

The green and the brown

All organic matter has a carbon-to-nitrogen ratio that expresses how fast or slow it will break down. All plant material is carbon-based; "greens" are low in carbon and break down quickly. "Browns" are high in carbon and break down more slowly. To get the most efficient compost going, mix materials to achieve a 30:1 ratio of carbon to nitrogen. If you combine fresh grass clippings and deciduous leaves, you will reach this ideal balance of materials for quick, efficient decomposition. Here are the ratios for common materials.

Yard waste composting

If you know a little bit about how composting works, then you can help the process along and make some high-quality compost to add to your garden. You will also be keeping organic matter out of the waste stream, saving the resources that it takes to haul and process this material.

A compost pile is a living thing; it needs air, food, and water. A healthy compost pile will support a vibrant community of beneficial bacteria and fungi as well as worms, isopods, springtails, millipedes, and other decomposers. Provide the proper habitat and these creatures will break down organic matter quickly into rich compost to improve your soil.

Good mix of materials

Give your compost pile a balanced diet of "greens" and "browns." Decomposers thrive in an environment where there is a mix of woody and herbaceous organic matter. If you have too many "greens," your pile will be slimy and foul-smelling. If you include mostly "browns," it will take a long time to decompose. Store fall leaves in a barrel to have a supply of "browns" ready to mix with your spring "greens."

Small pieces

The smaller the pieces are when you add them to your bin, the faster they will break down into compost. A large branch will break down but it will take a long time to become a tiny piece of compost. If the branch is chopped into many small pieces, there will be more surface area for the bacteria to work on and it will take less time for these little pieces to change into humus or compost. You may chop materials with a machete, a lawn mower, or a wood chipper/shredder.

Moisture

Without water, organic materials will not rot (or they will rot very slowly). Decomposers thrive in a moist environment. For quicker decomposition, make sure all your materials are as wet as a wrung-out sponge. Check your compost pile to make sure it remains moist; add more water if things dry out.

Air

Most decomposers are aerobic creatures like you and me—they thrive and multiply in an environment that is filled with air. As your pile settles, you may want to turn over the materials and fluff them to add the air these creatures need to do their important work.

Time

It takes time for plants to grow and time for them to break down to become part of the soil. Some very "green" materials may break down in two or three weeks. Most organics will take four to twelve months to break down. Be patient: Good things come to those who compost!

Compost bins

Many people wonder what kind of compost bin they should use. Bins don't do anything but keep your pile from taking over a big corner of your yard. A bin is not necessary, but it will keep your pile smaller and make managing your compost easier. Decide what kind of compost you are going to make.

Will you add yard debris as you weed throughout the season or will you make hot compost, which breaks down more quickly? You can choose a ready-made bin or build your own. If you will be making slow, passive compost, a holding bin with a large opening and lid will work well. If you want to make hot compost, look for a bin that has three fixed sides and a removable front for easy turning. There are many different compost bins on the market. Check with your city's solid waste department— many municipalities make subsidized compost bins available in the spring. Most bins that you can buy are designed to hold materials while they slowly decompose. These have a wide opening in the top and often have a little door in the bottom for harvesting the finished compost. To harvest or turn the material in these bins, you have to disassemble the bin.

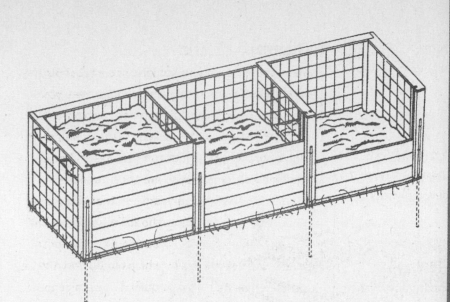

Make a Compost Sifter

Use some two-by-fours to build a box that fits across your wheelbarrow (or any other container that you will sift into). Cover the bottom of the box with half-inch square wire mesh or hardware cloth. Sift your homemade compost; use big pieces as mulch or put them back into your compost pile to rot some more. Use the sifted compost in your garden or containers.

Holding bins are not easy to use for hot composting. A homemade turning bin can be easily constructed out of recycled pallets or wood and wire. These bins have a bigger capacity and a side that opens or can be removed so material can be easily turned. Whatever bin you choose, make sure to use suitable organic materials, keep pieces small, and get everything as wet as a wrung-out sponge.

Where to put your compost bin

Put your compost bin in a convenient place with easy access to water and plenty of space for your wheelbarrow. You'll need some space to mix materials before you add them to your bin. Bins can be located in sun or shade; just make sure that your pile stays wet. Zone 2 is a great place for your compost bin. You may also want a space to store materials before you add them to your pile. Think about where you will keep your finished compost until you are ready to use it in the garden. Store finished compost in a place where it won't get wet—waterlogged compost is heavy and difficult to use.

Simple composting

Many of us add to our compost piles throughout the season as we weed and work in the garden. Make sure you use organic materials that are suitable for backyard composting. Add materials as you work in the yard. In 9 to 12 months, your compost will be done.

Tips for making great compost

SMALL PIECES

Make sure that the pieces are no longer than your hand and no bigger around than your thumb. Cut pieces small as you work so you aren't faced with a big disorganized pile at the end. If you wait, that unmanageable pile will likely sit there for weeks until, in frustration, you add it to the bin without being chopped or moistened. This won't lead to compost; rather, you'll end up with a dry, brushy pile that won't break down efficiently.

GET IT WET

A dry pile won't decompose very quickly. Speed up decomposition by making sure your materials are wet as a wrung-out sponge. When you are adding garden debris to the compost pile, make sure it is thoroughly wet. Find a partner and put all the debris on a tarp. One person can toss the material with a pitchfork while the other person waters it thoroughly with the hose. Mix the material well. Although you can get your compost too wet, typically piles are too dry. After the stuff is wet and mixed together, add it to your bin.

TROUBLESHOOTING FOR YOUR COMPOST PILE

SYMPTOM	CAUSE	REMEDY
Stinky, smells like garbage	Not enough air, too wet, too many "greens," food waste in bin	Turn pile to add air; add "brown" materials; remove food waste
Dry and doesn't break down	Pile is dry; may have too much woody material	Turn pile and add water; add more "green" material; chop woody material more finely
Wet but not breaking down	Not enough "greens"	Add "greens" or organic nitrogen fertilizer
Pile won't heat up	Pile is too small; not enough "greens"	Add more materials to fill a 3-cubic-foot bin; mix in more "green" material, chop pieces smaller
Slimy clumps of wet grass	Too many "greens"	Turn the pile to add air and break up chunks; mix in more brown leaves or straw
Pile looks like a brush heap	Too many large woody pieces	Shred or chop woody material; water thoroughly and add more "greens"
Pile is too hot— over 160°F	Too many "greens"	Add more "browns" and more water to cool the pile down

SIFT YOUR FINISHED COMPOST

At the end of the growing season, put the lid on your bin and let the creatures do their work. In the spring, when you open the bin to get some compost to add to your beds, the finished compost will be on the bottom. That's why many commercially manufactured bins have a little door at the bottom. Remove the material on top that hasn't composted and put it aside. You can get it wet and add it back into the bin to rot longer. The finished compost at the bottom will look like a mixture of soil, sticks, and twigs. This rough-looking material is finished compost, but it doesn't look like the stuff that comes out of the bags. To make your homemade compost look more like purchased compost, sift it through a half-inch screen. The large pieces that collect on top of the sifter can be remoistened and added back into the compost pile.

Advanced, hot composting

If you don't want to wait a whole year to harvest compost, mix "browns" and "greens" to make a hot compost pile that will be ready to use in a couple months. Hot compost piles are constructed so as to provide an environment that nurtures composting bacteria. These tiny, mighty creatures heat up the materials, making for an amazing science experiment and quick composting! To generate and hold the heat created by microbial activity, you will need enough organic material to make a three-cubic-foot pile. The amounts listed are only suggestions.

HERE IS A BASIC RECIPE FOR MAKING A HOT COMPOST PILE.

Ingredients

3 wheelbarrows of "greens," such as fresh grass clippings and small pieces of green plants

3 wheelbarrows of brown deciduous leaves

Water

A tarp to mix on and a pitchfork to toss material

Instructions

Mix half "green" and half "brown" in small batches and sprinkle with water as you toss it like a big salad. When the material is well mixed and wet as a wrung-out sponge, add it to the compost bin. Continue to mix "greens" and "browns" until all the material is in the bin. If you have the mix right, in three or four days the middle of the pile should heat up. Check the heat with a compost thermometer, which can be purchased at most garden centers. It should be around 140 degrees.

In 7 to 10 days, as the pile starts to cool down (and smell a bit like fresh manure), turn the material out onto a tarp, and toss well to add more air. Check the moisture and make sure that when you squeeze it you get a dribble of water. If the material is dry, add more water. Turn the pile every 10 days, for a total of three turns. Then let it cure—it should be ready to use in three to six months.

Food Waste Systems

Composting yard waste is a great way to keep and use the organic materials in your yard. After you consider the kinds of organic matter you have, you may find that you don't have much yard waste that is suitable for your compost pile. If you lack yard waste materials to build your soil, don't forget about the waste being created in your kitchen. Everyone makes food waste. Food scraps are sometimes a forgotten resource for making great compost.

If you will be composting your food waste, you will need a collection container to keep in your kitchen. There are lots of nifty collection buckets available. Look for a container that has an airtight lid to keep out fruit flies. The lid should be easy to remove with one hand since you will have food scraps in the other hand. Your collection bucket should have a ½- to 1½-gallon capacity. Keep your compost bucket on the counter or under the kitchen sink, close to where you will be creating food scraps. Empty and rinse out your bucket frequently to keep down smells and fruit flies.

Tips for Keeping Your Compost Bins Rodent-Free

- Keep a watchful eye.
- Visit your bins frequently, move material around, and add more water—this disrupts the habitat before rodents can settle in.
- Keep yard waste wet. Dry brush piles make wonderful warm, dry homes for rodents.
- Compost all food waste in a worm bin.
- Retrofit your worm bin so that it is more rodent-resistant. Raise the bin on blocks or bricks so that it is 12 to 18 inches above the ground. This way you can rake or sweep under it to keep the area clear of leaves and other debris.
- Line the outside bottom of your worm bin with 1/2 inch hardware cloth (square wire mesh) so that rodents don't gnaw the drainage holes bigger and enter the bin.
- Locate your compost bins 6 feet away from your fence line—rats don't like to cross open spaces.

What food can you compost?

Compost only vegetable food scraps—meat, dairy, and oils break down slowly and create smells that attract pests.

Two ways to compost food waste

FOOD WASTE BURIAL

The easiest way to compost your food waste is to bury it in an empty garden bed. This requires nothing but a shovel and an empty space in the garden. Burial saves you the time and energy of setting up bins, adding bedding, and harvesting. Just bury food scraps right where you will be planting.

Dig a hole about 12 inches deep and add your food scraps. Make sure to cover the food waste with at least 10 inches of soil. Pat the soil down firmly (but don't compact!) to discourage animals from digging. Check your burial site frequently to make sure there is no digging going on. In a month the food scraps will be gone, leaving a dark rich deposit of compost ready for planting.

Burial isn't a great option if you have dogs or cats who like to dig. Besides your own and local neighborhood dogs and cats, other critters—including raccoons, skunks, and opossums—are also known to dig for food. If you discover digging, try putting a large piece of wood or a garbage can lid, weighed down by a heavy stone or cinder block, over your burial site. If the digging continues, consider starting a worm bin.

WORM BINS

There is no better, easier, or more enriching way to compost food scraps than in a worm bin. Worm castings are considered "black gold" and are the Holy Grail of compost. Creating a worm bin allows you to experience the whole ecosystem of decay and renewal. Worm bins are different than yard waste composting because most of the decomposition is done by larger creatures (as well as microbes), most notably the mighty red wiggler worm.

OTHER FOOD WASTE COMPOSTERS

There are a number of other food waste compost contraptions on the market. They can be expensive and no more effective than burial or a worm bin. Read about them and check customer reviews before you make a big purchase. Beware of promises of ease and speed. It takes time for materials to break down and become soil.

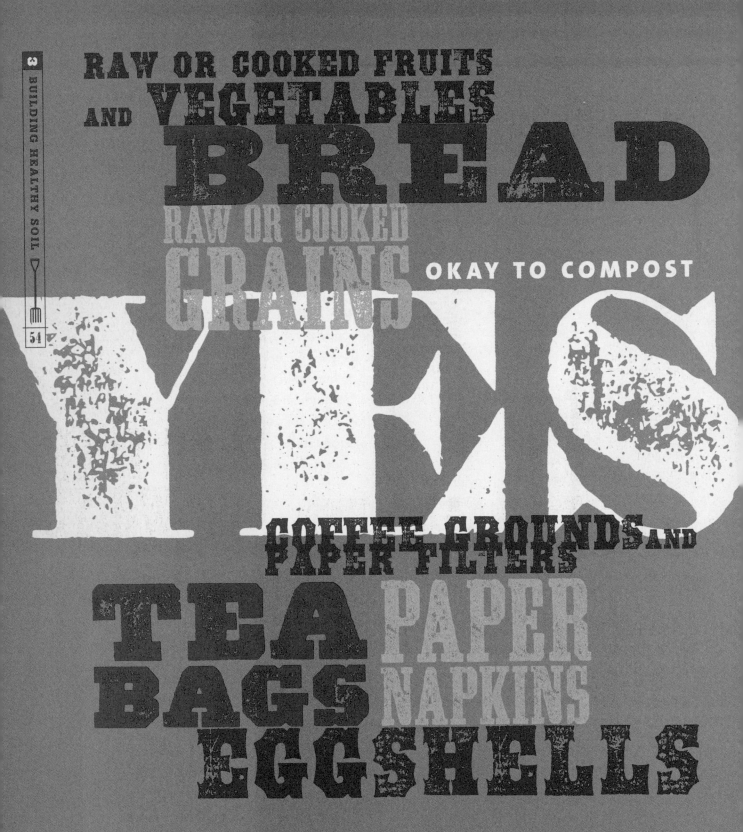

RAW OR COOKED FRUITS AND VEGETABLES BREAD RAW OR COOKED GRAINS

OKAY TO COMPOST

YES

COFFEE GROUNDS AND PAPER FILTERS

TEA BAGS PAPER NAPKINS

EGGSHELLS

Note: Food waste compost should be vegan.
Bread or cake with egg baked in it is fine.

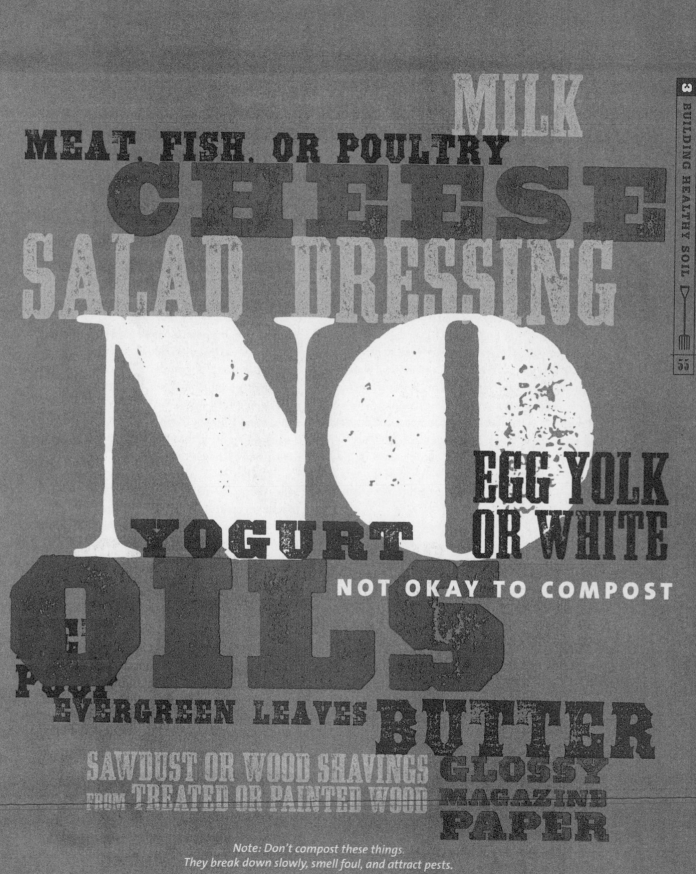

MILK

MEAT, FISH, OR POULTRY

CHEESE

SALAD DRESSING

NO

EGG YOLK OR WHITE

YOGURT

NOT OKAY TO COMPOST

OILS

EVERGREEN LEAVES

BUTTER

SAWDUST OR WOOD SHAVINGS FROM TREATED OR PAINTED WOOD

GLOSSY MAGAZINE PAPER

*Note: Don't compost these things.
They break down slowly, smell foul, and attract pests.*

Setting Up and Maintaining a
WORM BIN

A worm bin is a sturdy box where worms live and eat organic material, leaving their castings (or poop), which is excellent compost for your garden. As a rodent-resistant, closed environment, it is an excellent food waste composting option for city farms. A worm bin must be filled with a "brown" bedding material, which serves as the habitat for your compost critters. Your kitchen scraps are their food. A worm bin is also filled with a diverse community of decomposers, including red worms, sow bugs, centipedes, mites, springtails, ground beetles, and several gazillion microorganisms.

The Bin Itself

Before you start, you'll need to figure out how much food you will compost and how big a bin you'll need. Your bin should be large enough to handle all the material you want to compost. Worms can efficiently consume 1 pound of food waste per square foot of surface area each week. This is the rule of 1. A typical worm bin is 2 feet by 4 feet by 1 foot deep. This 8-square-foot bin can handle about 8 to 10 pounds of food waste each week.

Your bin will be a sturdy box made of plywood, heavy plastic, or metal with a tight-fitting lid so that larger creatures can't get inside. Drill drainage holes in the bottom of your bin: Worms can't live in standing water and food wastes are sloppy. If your worm bin will be inside or on concrete, place a drip pan underneath it to catch the goo. This watery compost extract or tea can be diluted 1:5 with water and used to water your garden or house plants—it is like a nutrient drink for your plants!

Put your worm bin in a protected area, close to where you will create food wastes. Avoid full sun, if possible—the worms don't like it too hot. A well-managed wooden worm bin will not smell and makes a great bench in an outdoor living area. Then you can surprise your guests by telling them that they are sitting on your worms!

Bed Your Bin

Bedding is the environment for worms and other decomposers. It is critical because without it you will have a stinky mess. Fill your bin with a combination of moist, shredded "browns," such as cardboard, brown leaves, sawdust, and newspaper. Make sure the bedding is wet as a wrung-out sponge. Fill the bin to the top!

Add Worms

Red wigglers are not the same as night crawlers or garden worms. Eisenia fetida, or red wiggler worms, are also known as manure worms. They are expert decomposers. They love a vegan diet and thrive in a wet, dark, temperate environment. Provide plenty of fresh bedding, along with your food scraps, and your population of worms will triple in a few months. No need to worry about overpopulation: Red wigglers stop breeding when they reach the carrying capacity of their environment. You can purchase red wiggler worms from local nurseries, bait shops, online, or get a starter batch from a friend with a worm bin. (See Resources for worms)

Feed Your Worms

Dig a hole or trench in the bedding and bury your food scraps under at least 6 inches of bedding—this will keep smells down and limit the breeding area for fruit flies. Stagger your burial across the bin using several quadrants. This gives the worms a few weeks to eat your food waste before you bury in the spot again. If you come back to an area and the food is not gone, you may need to cut back on how much you are feeding the bin. It may take three or four months for your worms to get established and start working their magic!

Tips for Maintaining Your Worm Bin

■ *To keep the bin materials moist, put a piece of plastic or flattened cardboard on top of the bedding. Incorporate the moist cardboard into the bedding as it gets too wet to lift off.*

■ *If the compost at the bottom of your bin is sloppy, add more absorbent bedding, such as coir (made from the husk of the coconut) or sawdust.*

■ *To prolong the life of your wooden worm bin, paint the outside with exterior latex paint and caulk around any joints to keep out rain.*

■ *Winterize your worm bin by filling it with plenty of fresh bedding in the fall. If you live in an area where the ground is frozen for several months, move your worm bin into a heated garage or shed. If your worms don't make it through the winter, buy a new batch in the spring, fluff the bedding, and start again.*

Harvesting and Rebedding Your Bin

Six to 12 months will pass before the bedding in your bin turns into compost. A new bin bedded with newspaper and cardboard may take several months to get going. To give your bin a jump-start, introduce additional microorganisms to your bin by adding a few handfuls of soil, leaves, or yard waste to your bedding. After a year, you can harvest your black gold. (Once your bin has been up and running for some time, you may harvest twice a year.) Finished compost will accumulate at the bottom of the bin. It will look like dark soil or coffee grounds. When the finished compost fills two-thirds of the bin, it is time to harvest.

The easiest way to harvest worm castings is to remove all but a third of the contents of the bin. Fill your box with fresh, moist bedding and continue burying your food waste. The worms, microbes, and compost left in your bin will keep everything going. You can sift the finished worm compost to remove any food that hasn't broken down and put it back in the bin to compost further. Incorporate both compost and worms into your garden beds.

Mulching

Leaves and branches drop around plants, creating a carpet of organic material that breaks down into humus and feeds plants. This carpet is mulch. You can imitate nature by adding mulch to your garden. Mulch is a layer of organic material, such as leaves, woodchips, grass clippings, or straw, which is spread out on the surface of the soil around plants. Mulch builds soil from the top down. Decomposers live in and eat the mulch and travel into the soil, mixing and loosening even highly compacted soils. Mulching is a great way to use organic material on your site to improve soil without the hassle of making a compost pile.

BENEFITS OF MULCHING

Protects roots

Builds soil by adding organic matter and nutrients

Eliminates weed and grass competition for space and water

Regulates soil temperature by insulating it from cold and heat

Provides habitat for worms, ground beetles, spiders, and beneficial fungi

Reduces soil compaction from heavy rainfall

Prevents erosion on sloped areas

How to mulch

■ *Weed the area you will be mulching.*

■ *Select organic materials that match the kind of plant you will be mulching—mulch like with like. For trees, shrubs, and woody plants, use woody mulch, such as woodchips, wood shavings, or leaves. For vegetables and flowers, use more herbaceous materials, such as straw, grass clippings, or leaves.*

■ *Spread mulch 2 to 6 inches deep around the plant, extending out to the outer edge of the branches or dripline.*

■ *Keep mulch away from the trunk or stem of plants. Leave a 3–6-inch mulch-free circle around the trunk or stem. Mulch that is pushed up to the trunk may allow diseases to enter the plant.*

■ *Apply mulch anytime during the season; a layer of mulch around plants will reduce the need for watering and weeding.*

Sheet-mulching

Sheet-mulching will change your life. This technique uses wet sheets of cardboard or newspaper and a super-thick layer of mulch to suffocate and kill unwanted grass and weeds. It enriches the soil, eliminates competition from other plants, brings in billions of decomposers, and helps increase water retention. Here's how it works: The wet cardboard kills the grass, and attracts worms and other decomposers. They eat the cardboard and mulch and, as the grass dies, they eat that, too. Composted turf makes some mighty fine soil! Sheet-mulching is a great way to create new garden beds, expand perennial beds, cover areas under trees, and make paths.

Sheet-mulching is easy—you just need a lot of cardboard, a pile of mulch, and a garden hose. Lay down flattened cardboard boxes, get them wet, and then layer some organic matter on top. Depending on what you are smothering and how thick your sheets are, you may need to do it again in a year or two. Sheet-mulching may not kill everything, but it will loosen the soil, making it easier to pull any weeds that burst through the layers.

How to sheet-mulch

■ *Measure your area and do some math to figure out how much you'll need for good coverage. Gather a lot of cardboard or newspaper. Find sheets that are not heavily covered with colored inks.*

■ *Get a big pile of mulch.*

■ *Remove tape or staples from the cardboard.*

■ *Cover the area in flattened cardboard or newspaper. Make sure the layers overlap and the ground doesn't show through. It is fine to have double or triple layers—the point is to smother the grass and weeds.*

■ *Use a sprinkler to get everything wet. Run the sprinkler for 10 or 15 minutes and then let everything rest for a while to absorb the water. Run the sprinkler again—it is good to get this layer nice and wet.*

■ *Spread a thick layer of mulch on the cardboard—12 inches or deeper.*

■ *Water the area with a sprinkler to give the microbes a boost and to help settle the mulch.*

SHEET-MULCHING MATERIALS

SHEETS

Flattened brown cardboard boxes *(avoid colored dyes)*

Newspaper *(without heavy colored-ink coverage)*

Burlap sacks

Cotton or wool rugs

MULCH

Woodchips *(don't use bark chips— see sidebar "Just Say No to Bark Chips")*

Straw

Deciduous leaves

Wood shavings

Pine needles

JUST SAY NO TO BARK CHIPS

The point of mulch is to provide organic matter that will hold water and provide a habitat for beneficial soil creatures. Bark chips are a popular mulch that might not be the right choice for you. Bark mulch is a by-product of the lumber industry. The bark is stripped off before logs are milled and then it's sold as landscape mulch. The bark on a tree keeps moisture and air from getting to the living tissue of the tree. It keeps the tree from rotting while it grows in one place for 200 years. When you spread bark over your planting beds, you are laying down an impenetrable layer that keeps out air, sheds water, and breaks down slowly. Use arborists' woodchips instead. Woodchips (which you can get from a tree service) include all parts of the plant. They break down more quickly, increase water retention and air space, and are an attractive mulch for your landscape.

■ *Spot-plant your new mulch bed by moving the mulch aside, cutting a hole through the cardboard, digging a hole, amending the soil and putting in your transplant. Instant landscaping!*

HINT: To keep grass from encroaching into your mulch, dig a shallow trench about 6 inches wide along the edge and remove sod or weeds. This is like a moat around your mulch bed to keep out invading grass!

Growing Your Own Fertilizer Using Cover Crops

Another important soil-building technique is to grow cover crops. These are plants that enrich the soil, support beneficial creatures, and provide organic matter. They are planted in empty beds at the end of the season to protect bare soil from the damaging effects of winter or harsh weather. In the spring, cover crops are cut into small pieces and then chopped into the soil to decompose. This biomass adds organic matter and replenishes the nutrients in your soil.

Cover crops can be used in the spring to enrich your soil until you are ready to plant. Grow a cover crop where you will put late-season or overwintering crops. Use perennial cover crops in paths and in future planting beds or in areas that need soil building.

Many different plants can be used as cover crops. Cereal grains have fibrous roots, which improve soil structure. Buckwheat and phacelia suffocate weeds, provide food for beneficial insects, and are easy to chop in. Legumes are important nitrogen-fixing crops (see sidebar "Mighty Nitrogen Fixers"). Growing a mix of grain and legume cover crops will fix nitrogen and build tilth.

Cover crop seeds are broadcast and watered just like any other seed. Three weeks before you want to plant or when 50 percent of the cover crop is flowering, cut plants into 6- to 8-inch pieces. Chop roots and plant material into your soil with a spade. Wait three to four weeks before planting to let organic matter decompose. For no-till gardening, cut plants into 4- to 6-inch pieces, spread plant material on top of the soil, and cover it with a burlap sack. Wait six to eight weeks for everything to decompose and then plant.

BENEFITS OF USING A COVER CROP

Fixes nitrogen

Fibrous roots improve soil structure or tilth

Attracts and supports beneficial insects

Adds organic matter, or biomass—roots, stems, leaves, and flowers that are incorporated into the soil

Protects soil from compaction—bare soil can suffer from heavy compaction due to winter and spring rain

Adds beauty and interest to your landscape

MIGHTY NITROGEN FIXERS

Growing legume cover crops is a great way to increase your soil's fertility. Rhizobacteria form a symbiotic relationship with legumes—peas, beans, vetch, and clover. These bacteria convert nitrogen from the air and feed it to the plant. In return, the plant feeds the bacteria carbohydrates. When rhizobia are active in your soil, you will see irregular pinkish nodules on the roots of your legumes. These nitrogen nodules when left to decompose in your soil will provide nitrogen for the next generation of plants. To take advantage of this—chop in legume roots and let them decompose before you plant.

Rhizobacteria occur naturally in your soil, but if you want to give your cover crops a boost, purchase some dried inoculant to use when you plant your seeds. To use inoculant, put seeds in a bowl, spray the seeds with a little diluted milk (this will help the inoculant powder to stick to your seeds), and then mix the seeds so that they are coated with inoculant. Plant them quickly, since the bacteria can't tolerate sunlight. Different strains of rhizobacteria work with different legume crops. When buying inoculants for your legumes, make sure you get the formula for the crops that you will be growing.

WHICH COVER CROPS ARE RIGHT FOR YOU?

There are many different plants that can be used as cover crops to build soil. Here are a few warm- and cold-season annual cover crops and two perennials.

		DESCRIPTION	SOIL REQUIREMENTS
WARM-WEATHER COVER CROPS	Buckwheat (*Fagopyrum esculentum*)	Does not fix nitrogen, but produces a lot of organic matter; fast-growing and good for smothering weeds. Can be chopped in after 6 weeks and then sown again: You can sow 3 to 4 buckwheat crops from May through August. Beneficial creatures love buckwheat flowers.	Likes fertile, well-drained soils; will tolerate light clay soils.
	Bush beans (*Phaseolus vulgaris*)	Rather than eat your beans, dig them into your garden beds. They fix nitrogen and provide organic matter.	Likes fertile, well-drained soils; will tolerate silty soils.
	Phacelia (*Phacelia tanacetifolia*)	Bees love this plant and its nectar-laden blue/purple flowers, a favorite of biodynamic farmers; cold hardiness to 25ºF; fast-growing, attractive plant; lots of biomass that is easy to chop in.	Will not grow in heavy or compacted soil.
	Red Cowpeas (*Vigna sinensis*)	These fix nitrogen, produce large biomass, and smother weeds. Fairly drought-tolerant after they are established.	Can grow in partial shade; likes well-drained sandy soil.
	Sunn hemp (*Crotolaria juncea*)	A vigorous legume that is great for fixing nitrogen and smothering weeds. Large taproot easily breaks up heavy soils.	Clay, highly compacted, or poor soil.
COLD-WEATHER COVER CROPS	Cereal rye (*Secale cereale*)	Cold-tolerant; will germinate down to 20ºF. Fibrous roots build soil structure; sow in fall with vetch.	Well-drained soil; can grow in partial shade.
	Common vetch (*Vicia sativa*)	Hardy to 0ºF. Produces large biomass and supports beneficial insects. Good at breaking up compacted soil; taproots can be 3–5' long. Sow with cereal grain to support vining growth.	Will grow in poor soils; fixes more nitrogen in fertile soils; tolerates some shade.
	Crimson clover (*Trifolium incarnatum*)	Popular cover crop that puts on a wonderful show of large crimson flowers. Easy to turn under, great as a groundcover around existing crops. Fixes nitrogen and attracts beneficial insects.	Tolerates a wide range of soils and does well in soil with low fertility.
	Fava beans (*Vicia faba*)	Beautiful plants produce large biomass, attract beneficial insects, and produce edible broad beans. Large taproot breaks up heavy soils.	Will grow in compacted or clay soils; can grow in partial shade.
	Winter wheat (*Triticum aestivum*)	Produces an extensive, fibrous root system that promotes the development of soil tilth (or structure); quick germinating.	Needs moderately fertile soil and full sun.
PERENNIAL	Alfalfa (*Medicago sativa*)	Fixes nitrogen and provides food for beneficial insects. Produces abundant biomass. Can be cut 2 or 3 times each season. Flowering tops can be added to compost or dug into poor soils.	Fertile, well-drained soil; full sun.
	Dutch white clover (*Trifolium repens*)	Low-growing clover (6–10"), fixes nitrogen, tolerates mowing and shade. Great cover for garden paths and under trees or vines. Flowering tops can be added to compost.	Wide range of soil types; thrives in wet, cool conditions.

Profile:
MASTER COMPOSTERS

Making compost is addictive. But good compost isn't quite as easy as tossing it all into a bin and hoping for the best: There is an art and a science to letting things rot. That's where Master Composters come in.

Master Composters are trained volunteers who help home gardeners set up compost sites and teach them how to build healthy soil. Seattle Tilth has been training MCs since the dawn of the worm bin! The Master Composter/Soil Builder program began in Seattle in 1986, and for more than 20 years, Seattle Tilth has trained these volunteers to teach others. Today, there are more than 500 Master Composter training programs in North America and abroad.

Each year, our program begins when Seattle Tilth selects 30 to 40 people from a competitive group of applicants and teaches them everything from the finer points of backyard composting to the nuts and bolts of industrial-scale organics recycling, and everything in between. Following their training, the volunteers are certified as Master Composters/Soil Builders. From there they move into the community to help others learn the wonders of composting. They support such projects as building a compost system at community gardens to designing and managing environmentally friendly waste-management plans for large events.

If you love compost and this kind of education and outreach sounds intriguing to you, check with your local solid waste management agency, cooperative extension service, or friendly environmental NGO to see if a program is offered in your area.

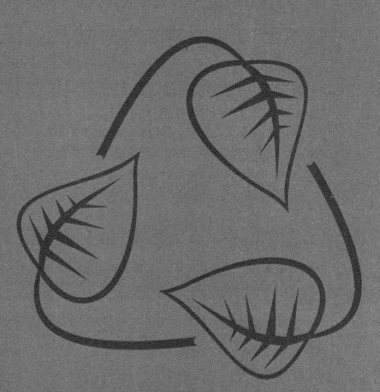

LEARN MORE:
seattletilth.org/
learn/mcsb

RAISED BEDS, CONTAINERS, & VERTICAL PLANTING

Now that you have assessed your site and know about your soil, it's time to make some beds.

IN THIS CHAPTER you'll learn different techniques for making garden beds in small urban spaces. You'll learn how to make raised beds, create straw bale beds, and do container gardening. It's time to get growing!

Where Will You Grow?

Many urban lots don't have a large, sunny, rectangular area for a garden. In urban settings you need to be creative and use lots of different small-space gardening techniques to grow the most food. Using a combination of raised beds, containers, and vertical growing spaces allows the urban farmer to utilize space most productively. What if the only really sunny spot is on a paved driveway? Containers with tomatoes, basil, peppers, eggplants, and climbing squash will allow you to produce food where you thought it was impossible.

Raised-Bed Gardening

One technique of small-space gardening is the use of raised beds that are heavily planted. Raised-bed gardening is a technique that has been embraced by urban farmers for years. These can be simple mounded beds or more extensively cultivated by double digging. Double digging is the process of removing 12 inches of soil and using a digging fork to cultivate deeply in the lower part of the soil. This takes a great deal of effort and with our abused and highly compacted urban soils, double digging isn't realistic for every site. Many urban farmers build their raised beds above the soil—building soil from the top down.

There are many benefits of raised-bed gardening. The soil in mounded beds heats up faster and holds heat better. Raised beds can be intensively planted, which means planting seeds or seedlings closer together than is usually

IS IT SAFE?

If you decide to frame your raised bed, don't use treated wood, railroad ties, or stained landscaping timbers for edible crops. The chemicals used to preserve wood may leach into the soil and be absorbed by the vegetables you are growing.

More worrisome is secondary contamination by accidentally ingesting chemicals from hands, clothing, and shoes that touch treated surfaces. Use only clean lumber or wood. A number of woods are naturally rot-resistant, such as cedar, oak, and cyprus.

recommended. This approach maximizes growing space. And since the garden bed is raised, there is more space for roots to go down (rather than out), which allows more plants to thrive in a small space. When plants are spaced close together, with leaves just touching, it shades the soil, which prevents weeds from germinating. Also, soil in raised beds tends to be loose, making them easier to weed and maintain.

Many city farmers like the look of a boxed, raised bed, which gives the vegetable garden a more formal look. Establishing and building a boxed raised bed takes a large initial investment. There is the cost of wood or the material that will frame the bed and the soil and compost to fill the box. Additionally, figure in the cost of replacing wood that rots—most wooden raised boxes will last only about five years, with repairs needed after the third year.

Thrifty, resourceful urban farmers can use old pallets, shipping boxes, old steamer trunks, and other wooden boxes made from clean lumber to make raised beds. Raised beds can also be framed with bricks, broken concrete, or cinder blocks. If your raised bed will sit on top of highly compacted soil, the minimum height is 12 inches. If your raised bed is sitting on a paved surface or rooftop, the minimum height is 24 inches.

Ripping Up Your Lawn and Making a Garden

The perfect place for your new garden may be covered by sod and weeds. You can dig up unwanted plants and compost them or you can sheet-mulch the area and rid yourself of unwanted plants through suffocation. One approach is to start small by digging one bed and sheet-mulching your future planting areas. Sheet-mulching (see chapter 3) takes time, but saves the labor of digging up and composting sod. After the sod breaks down, the mulched area will be a great garden bed. Sheet-mulch your new garden beds in summer or fall and then cultivate the following spring. Use deciduous leaves or straw as mulch for future veggie beds.

If you can't wait for sheet mulch, you'll have to dig up the sod and weeds. First, mark out the area. You can do this with bricks placed on the four corners or with stakes and surveyor's tape. Using a sharp spade or shovel, slice the sod into strips the width of your spade, then further into smaller pieces. If your soil is very hard, water the area and wait a day or two before digging. Push the shovel just below the grass rootline and pull back on the handle. This will lift up a wedge of grass and soil. Remove sod and shake off as much soil as possible. A compost sifter (see chapter 3) is handy for removing additional soil from roots.

After all the turf and sod has been removed, aerate your soil with a digging fork to break up any clods and remaining grass roots. Remove any big rocks. Amend your soil by adding 4 to 6 inches of compost. Sprinkle on a thin layer of all-purpose organic fertilizer (check the package and use half the recommended amount). Mix the compost and fertilizer into the top 6 to 8 inches of soil. Use a rake or your hands to smooth the soil so that it is flat. Your bed will be slightly raised. If you want the soil raised some more, bring in additional planting mix that can be purchased from a nursery or topsoil purveyor.

After you have created your raised bed, you will need to cultivate it each season before planting. Vegetable garden soil is compacted by rains, overhead irrigation, and gravity. Annual vegetables also lack the sturdy roots of perennial plants, which continually aerate the soil. Since repeated deep cultivation can destroy the structure of the soil, limit

Composting Sod

Composted turf is an excellent source of organic matter and mineral nutrients. To compost sod, first arrange pieces of sod—root side up—in a pile, watering well after each layer. This material should be wet as a wrung-out sponge. To speed up decomposition, add a nitrogen fertilizer, such as fresh chicken manure or alfalfa meal, to each layer. Cover with black plastic and let it sit for one to two years. After one year, dismantle the pile and sift the soil. Make a new pile with any sod that hasn't yet decomposed.

your cultivation to the top 3 to 6 inches. Shallow cultivation and mulching help develop good tilth or soil structure. Loosen the soil without turning it over and disrupting soil structure. Gently break up the soil enough to mix in compost or organic matter and to plant. Keeping soil covered with mulch or plants all the time will also help keep it from becoming compacted.

Too wet to dig?

Make sure your soil isn't too wet when you start digging in the spring. If your soil is too wet when you dig, it breaks into clods. These clods remain intact when the soil dries and can be hard as rock.

Here's an easy test for soil moisture:

1. *Grab a handful of soil from your garden.*

2. *Squeeze it into a loose ball (just so it holds together).*

3. *Throw this ball into the air about 8" above your hand.*

4. *Let the ball fall and hit your hand.*

5. *If the ball breaks apart easily, you can start digging. If it stays together, the soil is too wet and you'll need to wait.*

If your soil is too wet to cultivate, try covering your beds with burlap sacks and clear plastic a few weeks before you want to plant. The plastic keeps out the rain and the burlap brings worms and decomposers to the surface. Remove the plastic and burlap when the soil is just right to start digging.

Straw-Bale Gardening

Straw-bale gardening is economical, doesn't require any tools, is great for soil building, and uses a fraction of the soil needed to fill a raised bed or container. You basically build raised beds on top of bales of straw. If your garden has compacted soil or very poor drainage, this technique allows you to plant right away and build soil at the same time.

You'll need to purchase a couple bales of straw (not hay), a few cubic feet of potting soil, and some nitrogen fertilizer or fresh manure. The straw bale is meant to decompose in place, building the soil from the top down. If you choose to leave the strings

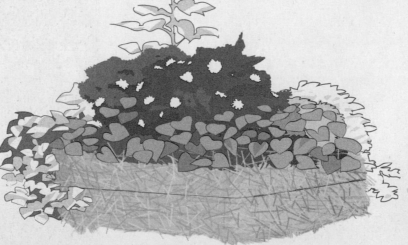

on, set the bale on its side. The cut ends of the straw will be pointing up and the string will wrap around the sides of the bale. This makes a tall, narrow bed. Place two bales side by side or make a bigger four-bale bed. If you choose to dismantle the bale, stack the straw into your desired shape and add soil layers.

Put a small sprinkler on top of the straw and run it just long enough so that water soaks the bale thoroughly. Add high-nitrogen fertilizer or manure. Water again well. This will help the straw decompose.

Add 2 inches of compost and then 6 to 8 inches of lightly moistened potting soil to the top of the bale. Your goal is to make a broad, flat gardening space. If your soil is too dry, it will fall off the sides of the bales and leave you with a narrow strip as a planting area. Pack the sides down as you build up the layers of soil. Try not to pack the middle, which is the planting zone. Mix in some organic fertilizer. Form a nice level mesa, then plant!

Container Gardening

Growing edibles in containers is easy, and adding containers to your city farm increases the flexibility of the growing spaces in your yard. Container gardening is a great option if you live in an apartment, are renting, or have no inground space in which to grow vegetables. Including containers in your garden plan

Easy Edibles
for
Containers

You can grow just about anything in a container. Depending on what you choose, you may need a really big container. Fruit trees, vegetables, herbs, flowers, vines, and fruit shrubs all grow well in containers. Select dwarf, bush, or "compact" varieties, which are well-suited for growing in the confines of a large pot.

ANNUAL VEGGIES AND HERBS
Sun—*tomatoes, eggplant, squash, cucumbers, beans, onions, leeks, peppers, basil* Partial shade—*salad greens, cilantro*

ANNUAL EDIBLE FLOWERS
Mix edible flowers with herbs in your containers.

Sun—*sunflower, marigold, bachelor's buttons, nasturtium* Sun, partial shade—*calendula, borage, pansy, salad burnett* Shade, partial sun—*begonias*

PERENNIALS: BERRIES, FRUIT AND HERBS
Sun—artichoke, dwarf fruit trees, blueberry, raspberries, strawberries, rosemary, sage, parsley, thyme, oregano Part-shade—raspberry, blackberry, boysenberry, alpine strawberry, evergreen huckleberry, mint, sorrel, comfrey

is a great option for yards that are shady and where the only sunny spots aren't near soil. Containers filled with vegetables can also be integrated into existing plantings or located where soil is too compacted or contaminated.

Containers add a wonderful accent to garden beds and offer a terrific way to include invasive edibles, such as mint or comfrey, in your garden plan. They are easy to maintain and are virtually weedless. They can be located on sidewalks, driveways, decks, patios, rooftops, and windowsills, at the ends of garden beds, or integrated into landscape plantings.

Choosing a container

Containers can be made out of almost anything. Any nontoxic container more than 12 inches deep with drainage holes is a potential raised garden. The bigger the container, the more you can grow. Dark colors will conduct more heat, which is great

VEGETABLE/HERB	VARIETIES OR TRAITS TO LOOK FOR	MINIMUM CONTAINER SIZE
Beans	Any bush variety	3–5-gallon container
Beets	All kinds; harvest when small for baby beets	Wide container, 14–24" deep
Carrots	Romeo, Thumbelina, or other round varieties	Wide container, 14–24" deep
Cucumber	Any bush or semibush variety	Shallow, wide 3–5-gallon container
Edible flowers	Calendula, marigolds, petunias, nasturtium, pansies, alyssum and violas	8–12" deep or deeper
Eggplant	All kinds with smaller fruit	3–5-gallon container
Greens	Kale, chard, lettuce, spinach, mustard greens, bok choy, radicchio, and arugula	Window boxes or any container 8–12" deep
Onions	All types	10–12" deep or deeper
Peppers	Any sweet or hot pepper variety	3-gallon container
Radish	Round varieties	Wide container, 12–18" deep
Strawberries	Any kind	Strawberry pot or any container at least 8–10" deep
Summer squash	Bush varieties; climbing varieties in a very large pot	5-gallon container; 15–20-gallon for climbing squash
Tomatoes	Determinate or bush varieties	5–10-gallon container
Annual herbs	Basil, cilantro, dill, chamomile, chervil, lemongrass, shiso	10" deep or more
Perennial herbs	Rosemary, thyme, hyssop, sage, lavender, mint, oregano, marjoram, catnip, verbena	The larger the container, the better they will grow. Try a 5-gallon container planted with 2 or 3 different herbs

for heat-loving plants, such as tomatoes, basil, and peppers. However, heat may be injurious to young, tender seedlings. Porous containers, such as clay, unglazed pottery, wood, and concrete, will dry out faster than plastic.

Materials

- *Plastic pots are light and hold water well, come in many colors, and are inexpensive. Plastic can fade or crack in sun over time.*

- *Terra-cotta dries out very quickly and requires more attention to water properly. Terra-cotta pots are inexpensive but will crack if left out in freezing temperatures.*

- *Glazed-clay pots hold moisture better than terra-cotta and come in many colors but can be pricey.*

- *Rice or sunflower hull recycled pots are a new biodegradable type of container that are sturdy, lightweight and look nice. They last three to four seasons before they start to degrade.*

- *Fiber-resin containers are made from polyethylene plastic, but are made to look like stone, clay, or wood. They are inexpensive, very lightweight, but may crack with sun exposure.*

- *Half-whiskey barrels or wine casks are recycled containers 2 feet in diameter and over a foot deep, perfect for growing vegetables or herbs. Drill four large drainage holes in the bottom. Half-barrels last three seasons or longer before wood starts to rot.*

- *Get creative! Plant in nursery pots, shipping crates, kiddie pools, 5-gallon buckets, plastic storage tubs, or burlap sacks.*

How to plant

Before you fill your container with soil, put it where you want it to be. Large pots filled with soil are heavy and very hard to move! If the container will be on soil, weed the area and put down a weed barrier (such as a burlap sack, cardboard, or newspaper) then install your container. Containers can stain concrete or wood decking, so use a drip tray to protect surfaces.

Before filling your container, empty your potting soil into a wheelbarrow or a large tub. Work water into the potting soil with a trowel or with your hands. The planting mix should be moist but not soggy, loose and easy to work with.

Fill the container loosely with potting soil. Leave about 2 inches of space at the top of your container. Shake the container to settle the soil. Don't put stones in the bottom; this does not improve drainage. If the hole in the bottom of your container is large, cover it with a piece of mesh screen or porous fabric to keep the soil from falling out.

Now you're ready to plant!

BASIC POTTING SOIL

1 part coco coir

1 part perlite

1 part sand or pumice

1 part compost

Source: MNGG

Potting soils

Potting soils are specially formulated for good drainage, organic matter, and water retention. Avoid potting soils with added fertilizer—you will want to know what you are putting into your mix.

Fertilization

Containers need more fertilization than in-ground garden beds. The population of microbes is smaller and roots can't mine soil for nutrients. Mix in a slow release, all-purpose organic fertilizer at planting time and then use liquid fertilizer during growing

ACCESSIBLE GARDENING

Everyone can garden. With thoughtful garden-bed design and placement, those with physical limitations can grow food for the table too. Keep gardening accessible for all by using the following techniques:

■ Build raised-bed planters at a height of at least 18 inches. In a wheelchair or seated on a bench or chair, 3 feet may be more comfortable. This allows you to sit down and garden. Build beds no more than 2 feet wide so that you can reach all the way across from one side.

■ Bench seats can be built into the sides of beds.

■ Tabletop raised beds with clearance underneath will allow wheelchairs to slip under the planter and let the gardener use both hands, instead of parking a wheelchair alongside the bed.

■ A full watering can may be too heavy to lift and pour from, so consider alternatives, such as small cans made of lightweight material or use a hose extender (just a short piece of hose) and set up a faucet at the edge of the raised bed, within easy reach. The water can be left on at the original faucet, and turned on and off at the raised-bed site. The control for a drip irrigation system can be set up at the edge of the raised bed.

■ Include a flat surface to act as a small table so that hand tools and harvest baskets are within easy reach.

■ Paths from the house to the garden should be wide enough to accommodate a wheelchair. Americans with Disabilities Act (ADA) requirements say that 32 inches is wide enough for a doorway, but consider allowing 40 inches to give yourself extra turning room and extra width to accommodate tool handles and harvest baskets.

■ Create a smooth hard surface for wheelchairs to roll on—this is no place for 6 inches of pea gravel.

■ Consider including weatherproof containers near the beds for fertilizer and hand tools, so that everything you need will be handy.

■ When choosing what to grow, don't forget fruit trees—apples grown on mini-dwarf rootstocks can be kept 4 feet high and trained on wires for easy access.

■ Adapted hand tools help reduce stress on your wrist, and some are designed for those with reduced finger mobility and dexterity. A Web search will help you find sources.

season. Liquid fertilizers are bottled concentrates that are diluted with water. Liquid fertilizers provide water-soluble nutrients that are readily available to your plants and don't require soil organisms to convert them. Use liquid fertilizer every two to three weeks for big containers (5 gallons or bigger), every one to two weeks for smaller containers (less than 5 gallons).

Organic Liquid Fertilizers for Edibles

- *Fox Farm Tiger Bloom or Grow Big Organic Liquid Fertilizer*
- *Earth Juice Bloom or Grow Fertilizer*
- *Alaska Fish Fertilizer*
- *Age Old Organics Liquid Bloom and Grow*

Watering

Water is essential for healthy plants. Vegetables need a lot of water. You will water your containers more frequently than in-ground beds, because roots can't reach down for the moisture in the earth. Be sure not to let your pots dry out. Check the soil for moisture by digging into the soil 2 inches to see if it feels wet. Experiment with how much and how often you need to water each container to keep the soil moist. In really hot spots, and with smaller containers, you may need to water more than once a day.

Keeping it growing

Growing clover or mulching with leaves or grass clippings between vegetable crops will help maintain the vitality of the soil in your containers. You may need to add more compost and fertilizer to older potting soil. Rejuvenate your pots each season by removing one-third of the soil and adding half new potting soil and half compost. If plant vigor and production decreases, replace all the soil in your container. Rotate your crops as you would with your other garden beds to prevent pests and diseases. If you experience disease problems in your containers, discard the soil and clean the container with a solution of 1 tablespoon of bleach in 1 gallon of water.

Small-Space Techniques

Using a variety of gardening techniques makes the most out of a small space. Options include growing vertically, planting crops in succession, and interplanting to produce more food on your city farm.

Vertical growing

If you only have room for only a small patch of soil, your best option may be a vertical garden. Creating trellises, tepees, and other arbors can increase space in small gardens. Plants that are growing up a trellis have more exposure to sun and wind. Great plants for vertical growing are cucumbers, tomatoes, peas, pole beans, runner beans, hops, vining berries, jack-be-little pumpkins, gourds, climbing summer squash (trombocino), winter squash, sweet potatoes, and pumpkins (these may need nylon slings to support the weight of plump heavy fruit).

Trellises and arbors are wonderful features for the garden. They add interest and height in what may otherwise be a single layer of plants. Harvesting from a trellis is a breeze. Rather than the bent-over, backbreaking work of harvesting bush beans, pole beans grow at a convenient height for picking. Tomatoes love a structure so they can stretch their long, pungent vines out with easy-picking fruits dangling from trellis strings. Peas and pole beans would be nearly impossible to harvest if they were not grown on a trellis.

When you build your trellis, think of it as a semi-permanent structure. It will be taken down at the end of the season, but will need to stay in place for four to six months. Your trellis must be strong enough to bear the weight of mature plants and the stress of fall winds and rain. If you are growing scarlet runner beans, for example, your beans will be on the trellis until just before your first frost or the rainy season commences. If your trellis falls over, it will be difficult to stand it up again. Your plants may not survive the stress and strain of reinstallation.

Construct your trellis before the plants need the support. That way the trellis is ready as plants start sending up snaking vines or reaching out with sticky tendrils. Putting up your trellis before you need it will prevent your plants from forming a tangled mass that will be difficult to train on a structure. Build your trellis on the ground first. Before installing your trellis in the garden, put on all the strings and attach all the crossbars. It's easier and safer than climbing a ladder or balancing on a bucket to finish the job.

Trellises may be made out of a variety of materials, including bamboo, cedar stakes, straight branches, or metal conduit pipe. Inexpensive jute twine is strong enough to carry the weight of most plants during the growing season. Since it decomposes, it can be ripped easily from trellis poles and added to the compost pile. Use heavy cotton clothesline to secure the top of a tepee, crossbars, and other supports that will bear the weight of plants. You can use old fencing or chicken wire for your trellis. Use chicken wire to form a small pyramid or tepee around the waist of a scarecrow extending all the way to the ground. Now plant cucumber or trailing nasturtium seeds around the bottom and watch the topiary skirt grow! Have fun!

Snakes and rock climbers

Not all vining plants need the same kind of trellis. Some need vertical lines and others require horizontal strings. Pole and runner beans are like snakes—they need verticals to wind around. Beans planted around a bamboo tepee will twine up strings and poles at breakneck speed. Put one or two vertical strings between each pole and attach each one to a horizontal line tied around the base. Leave a space between the legs as an entrance; tie another horizontal line 3 or 4 feet up to keep the plants from draping into the center of your tepee.

Peas and any climbing squash—plants that have those little whiplike tendrils—are like rock climbers (or Spider-Man) and need horizontals to cling to. It is easy to set two bamboo poles in the ground and then weave a net with clothesline and jute twine for cucumbers or peas to climb on. Space horizontals 2 to 3 inches apart, increasing the spacing as you build up. Add vertical lines to create stability and to wrap stems around.

Check the seed packet for how tall your variety will grow and build your trellis a bit taller. Runner beans will grow to more than 15 feet and some snap peas will crest at 7 feet. Building a tepee that is too short is almost as bad as having a trellis that falls over!

Edible food fort for kids

Create a special space for kids in the veggie garden. Making trellises that support vining plants and double as secret clubhouses are great family projects that add to the magic of the garden for children. Bean tepees are excellent hiding spots from which to spy on family members and munch on tasty garden snacks. Spreading out with books and crayons in the shade of a scarlet runner bean tepee is the stuff of lasting childhood memories.

After you decide where the tepee will go, figure out what

MAKE GARDEN MEMORIES

HOW TO BUILD A
BEAN TEPEE

MATERIALS YOU'LL NEED

4 or more straight poles—
10–15' long
Strong cotton rope
(clothesline works well)
1 or 2 balls of jute twine
Shovel
Golfball-sized rocks
Runner or pole bean seeds

INSTRUCTIONS

1. Line up three or more long poles together on the ground. About a foot from the top end, wrap them together with the cotton rope or clothesline. After looping the rope two or three times, tie it loosely (your hand should be able to slip into the loop).

2. Use another section of cord to loop around the first cord between each pole, weaving in and around each pole until the first cord is taut.

3. Add 8–16 vertical jute lines and tie them to the top of the structure while it is still on the ground. These will be the lines between each of the legs.

4. When you have added all the vertical lines, stand it up, splay the legs out, and position them where you want the outside perimeter of the tepee. A tepee for one or two children should be about 4 feet in diameter— large enough to crawl in yet small enough to provide an excellent hiding place.

5. Dig small, deep (12-inch) holes for each leg pole.

6. Put each leg into the hole and add a few rocks around the each leg to give it stability, then fill it about half-full with dirt and tamp down with a heavy wooden handle (as you would when sinking a fence post). Add more dirt and tamp down more until the holes are full and the tepee legs are solid in the ground. Remember that this structure will be carrying a lot of weight after all the bean vines and fruit cover the frame. Think of this as a semipermanent structure.

7. Spread out the vertical lines so that they are evenly spaced around the tepee.

8. Tie a horizontal piece of twine around each leg pole at intervals about a foot above the ground. Tie the vertical twine to this horizontal and leave a tail to the ground for young beans to meander up. Leave one or two openings between the legs so you can enter and exit the tepee.

9. Add two or three more horizontal pieces of twine, spaced 3–4 feet apart, and going up as high as is comfortable to reach. These horizontal lines will keep the plants from draping into the center. Beans climb like snakes, winding around vertical lines and poles, so these horizontal pieces of twine help the tepee keep its shape.

10. Plant bean seeds on the outside perimeter around and between each pole leg.

kind of beans you want to eat as well as hide among. Runner beans generally produce dry beans; only the very smallest pods are worth eating. The surprise edible with runners are the delicious flowers. These scarlet blossoms are sweet, crunchy, and beany—like the sweetest of all green beans! Runner bean vines populate fairy tales and grow to 15 feet. Pole beans make shorter, 8–10-foot vines and come in dry, wax, and snap types. For classic green beans, look for a snap bean. Blue Lake and Kentucky Blue Wonder are sturdy varieties for any edible fort.

Succession planting

Succession planting is a method of sowing staggered plantings so that crops don't all ripen at once. For a continuous harvest throughout the season, plant seeds or starts at intervals of every few weeks. Any crop that you eat fresh is a great candidate for succession planting. You may have one or two successions of beans or you may plant every few weeks for baby salad greens.

Watching over so many seeds in so many different stages can be challenging. Year-round gardening is another succession planting technique you can try. Plant different crops spring, summer, and fall in the same garden bed. Fill the spots that have been vacated by early crops with the next round of plants. Cool-season crops, such as spinach or lettuce, can be grown in the spring followed by

THEME GARDENS

When space it limited, consider growing the ingredients for your favorite dishes. Many urban farms include theme gardens or beds. Growing the three sisters—squash, beans, and corn—is a common theme garden. Try a Russian heirloom tomato bed. Plant a nacho garden with tomatillos, peppers, green onions, and cilantro for homemade salsa verde. Grow herbs, tomatoes, and wheat for homegrown pizza—sure to be the highlight of the farming season for your family!

Fancy gourmet salad gardens allow you to grow your favorite lettuces while adding color to the garden. Pick herbs and Meyer lemons grown in containers, and you have salad and dressing just beyond your doorstep. When the vegetable garden seems monochromatic green, include rainbow vegetables that provide brilliant splashes of color. Bright lights chard, speckled trout lettuce, purple tomatillos, and a variety of edible cutting flowers are a feast for the eye and delicious on your plate.

a warm-season crop of beans or squash with a fall planting of corn salad or a cover crop. Time it right and you can eat from your yard most of the year!

Succession harvesting is a fun family activity. Beets and carrots love to be sown thickly, which means they need to be thinned. Thinning beets or carrots is essential so that there is space for roots to get big. Eating the thinnings allows you to enjoy baby beets and greens before you finally harvest full-sized roots.

Interplanting

Interplanting is a technique of planting more than one type of vegetable in the same garden bed or container. This is a great way to grow a diversity of crops in a small space. Arrange plants by sun, soil, and water needs. Common combinations include mixing tall and short plants; cool- and warm-season plants; undersowing cover crops; and mixing slow- and quick-growing plants.

TALL AND SHORT PLANTS

Mix tall plants, or vertical growers, with shorter vegetables and flowers. Plant a row of pole beans behind a border of edible flowers, such as pansies, calendula, and nasturtiums. When in full bloom, this creates a curtain of color covered with tasty flowers and crunchy pods.

QUICK- AND SLOW-GROWING PLANTS

Plant a quick-growing crop among or behind a long-season crop.

Sow cilantro seeds behind your leek starts. The leek starts are tiny and will not shade out the cilantro. You may cut and have cilantro come again two or three times before it flowers. The leeks will still be fairly small when it is time to pull up the cilantro.

COOL-SEASON AND WARM-WEATHER CROPS

Plant lettuce starts spaced 8 to 10 inches apart, then sow bush beans in the empty spaces. The beans will germinate while the lettuce is still small and will shade the lettuce so that it will be slower to bolt.

UNDERSOW WITH CLOVER

After plants are established, sprinkle annual clover seed in the spaces between and under the plants. This low-growing legume will build the soil, fix nitrogen, outcompete weeds, and prevent compaction. When vegetable plants are removed, your cover crop is already filling the space. Chop in your clover crop three to four weeks before you want to plant your next crop.

Integrating Edibles into Your Landscape

Now that you know all about the different techniques for small-space growing, integrating edibles into your existing landscape is a snap! You can grow vegetables and fruit between your existing trees and shrubs or replace ornamentals with something you can eat. Make sure the conditions are right before you plant. Look for spots that

edible LANDSCAPE PLANTS

TREES: dwarf apple, pear, dwarf cherry, plum, serviceberry, fig

SHRUBS: strawberry, blueberry, evergreen huckleberry, black elderberry

VINES AND CANES: grapes, hardy kiwi, passionfruit, hops, thornless blackberry

HERBS: rosemary, lavender, thyme, sage, dill, chives, tarragon, mint, parsley, basil, oregano, marjoram, chamomile, cilantro

PERENNIAL AND BIENNIAL VEGETABLES: bulbing fennel, rhubarb, artichoke, cardoon, asparagus, sorrel, lovage, leeks, garlic, shallots, onion

ANNUAL VEGETABLES: lettuce, spinach, bok choy, tomatoes, peppers, tomatillos, celery, carrots, beets, turnips, peas, beans, parsnips, shiso, cabbage, collards, broccoli, purple cauliflower, lacinato kale, radicchio, mustards, rainbow chard, red kale

EDIBLE FLOWERS: viola, borage, nasturtium, calendula, pansies, dianthus, bachelor's buttons

get at least six hours of sun and are easy to access for watering and harvesting. There are three ways to add edibles to your landscape: sheet-mulch areas to make new planting beds; incorporate containers where you can't dig; and interplant vegetables and edible flowers among your perennial plants.

Sheet-mulching under Trees

Enlarge existing beds or sheet-mulch under trees to create new planting areas for edibles. Lay down cardboard and spread mulch to the dripline, which is the outermost edge of the tree canopy. Prune lower limbs of trees and shrubs up to bring in more light. Spot-plant in the mulch by pushing mulch aside and cutting a hole in the cardboard. Dig a hole

slightly larger than your transplant. Add a couple inches of compost and a little fertilizer to the hole, then mix well before transplanting your veggie start.

Interplanting among your Existing Plants

If there are open spots in your planting beds, slip in something you can eat. Onions, lettuce, pansies, calendula, and Swiss chard all have shallow root systems and will grow in soil shared by ornamentals. Soil in perennial beds will need to be amended before you plant. Loosen soil and mix in a few inches of compost and some fertilizer, then plant seeds or starts. Start small by planting a few green onions, herbs, or chard between perennials in sunny spots.

Containers Everywhere

Plant in containers if the roots of perennials or trees are shallow and it is impossible to cultivate the soil. A group of containers filled with vegetables and edible flowers on the south side of an arborvitae hedge can provide beauty and food in an otherwise barren urban landscape. Planting pole beans, letting them twine up a tall hedge, and mixing in trailing nasturtiums in containers delights the eye as well as the palate.

Path to Freedom

Using a combination of raised beds, containers and planting edibles among your existing landscape is a great way to grow lots of food in a small space. Breaking out of traditional rows and spreading food crops around your yard creates a beautiful, multipurpose environment for your family to enjoy.

Profile: URBAN HOMESTEAD

LIVING OFF THE LAND—it's a dream of many gardeners. But usually the dream includes moving to the country, where there's room for a big garden with a barn and a pasture for the animals. How big is this dream garden—5 acres? 10 acres? 50?

How about 1/10th of an acre, the size of a regular neighborhood lot in the city? That's the challenge accepted by the Dervaes family, who began homesteading on 10 acres in New Zealand and ended up, in the mid-1990s, homesteading on a tiny lot in Pasadena, California.

Today, the Dervaes grow 6,000 pounds of food a year, enough for themselves and to sell to local restaurants. Their garden produces a constant supply of a wide variety of fruits and vegetables. They also raise chickens, rabbits, and goats.

Conservation is a cornerstone in the family's lifestyle. Solar panels produce much of their electricity needs; the yard and house are intensely gardened ("square-inch gardening" they call it); organic matter is recycled back into the garden; they save their own seeds; keep bees and sell the honey—and that's just a small fraction of what they do.

Their high-yield practices have led to lots of interest, and so the Dervaes family has created a highly informative website. They also sell their own seeds and offer many other self-sufficiency products for sale. Read about the family and what they've accomplished—you'll be inspired to do more for yourself.

LEARN MORE:
urbanhomestead.org

STARTING WITH SEEDS

In this chapter, you will learn how to select seeds and starts. You'll learn how to plant outside, how to grow your own transplants indoors, and how to save seed.

Growing from Seeds or Starts

Great gardens start with healthy seeds or plant starts either purchased from a local nursery or that you have grown yourself. Look for vegetable and herb starts at nurseries, farmers' markets, or local plant sales (these are often fund-raisers for worthy causes). Seeds can be purchased at nurseries, at garden stores, or by mail order. Ordering seeds by mail is relaxing and can be done from the comfort of your home.

BUSH BEANS
REFUGEE
CARD SEED. C
FREDONIA, N.Y.

A BACHELOR'S BUTTON
PINKIE
F. LAGOMARSINO & SONS
SACRAMENTO, CALIFORNIA

RADISH
EARLY RED TURNIP

PEAS

Seeds or starts?

There are many advantages to growing your garden from plant starts. Vegetable transplants give you a jump-start on the gardening season. Transplants are already a couple of months old so the time to harvest is shortened. It is easier to distinguish a vegetable transplant from a weed and easier for beginning gardeners to correctly space plants. Seeds are very small and it is sometimes difficult to remember that they will grow into big plants. Starts are already plants, so it is easier to visualize them growing and taking up space.

Planting seeds in pots or in the ground is easy and a great activity for children. Seeds are inexpensive, wondrous, and beautiful. There is nothing like watching a tiny sprout burst through the soil and develop stems and leaves. You may wait a bit longer for your harvest from seeds. Growing your own seedlings is economical and a great way to get a jump on the season.

Many vegetables are better grown from seed because they don't transplant well. Carrots, beets, corn, beans, cucumbers, and squash should all be sown directly in the garden because transplanting disrupts root formation.

Selecting Varieties

Whether you are growing from seeds or starts, pick varieties and crops that will be easy to grow and produce well for you. Find varieties that do well in your climate. Look for fresh seed, grow short-season crops, and plant a mix of cool- and warm-weather crops. Purchase seed that can be saved to plant again.

Easy to grow

Choosing vegetable varieties that grow well in your climate will make your job as a gardener easier. Favorites, such as beans, peas, lettuce, and kale, are a snap to grow, provided you plant them at the right time. If you will be using containers, look for compact or climbing varieties that are better suited to a small growing environment.

Choose fresh seed

Look at the "packaged-for" date. This will ensure that you are buying fresh seeds, which will have a high germination rate. Most seeds will last two to three years if stored in a cool, dry dark place.

BEST GROWN FROM SEED
Carrots, beets, turnips, radishes, beans, peas, corn, squash, pumpkins, cucumbers, spinach, cilantro

BEST GROWN FROM STARTS
Tomatoes, tomatillos, peppers, eggplant, herbs, leeks, onions, lettuce, broccoli, cauliflower, cabbage

Shorter-season crops

Select short-season crops by looking for the days-to-maturity on the seed packet. This is the number of days from transplanting or direct-sowing to when you might expect to harvest. Short-season crops are those that require 65 days or fewer to maturity. Short-season crops are more pest-resistant. Since they are in the ground for a shorter period, there is less time for bugs or disease to attack. Quick crops make it possible to grow more than one crop each season in your garden spaces.

What to Look for When Purchasing Seeds

Packaged-for date

Open-pollinated, organic, rare, endangered, or heirloom seeds

Varieties that do well in your climate

Compact varieties for small spaces

Days-to-maturity—look for short-season varieties, 65 days or fewer

What to Look for When Buying Plant Starts

Healthy and free of disease and insects

Sturdy, compact plants— not leggy or fragile

Roots that look healthy and white and are not overly crowded in the pot

COOL- AND WARM-SEASON CROPS

COOL
beets
cabbage tribe
cilantro
lettuce
peas
spinach
Swiss chard

WARM
beans
corn
eggplant
peppers
squash
tomatillos
tomatoes

Cool- or warm-season crops

Your garden vegetables will not all be planted at the same time. Some vegetables grow in the cool season (spring and fall) while heat-loving crops are planted later (late spring, summer). Some vegetables, like broccoli, will grow in both cool and warm seasons as long as you select the right seed variety. Crops grown out of season will be more susceptible to pests and disease and will fail to thrive.

Seed choices

Those who own the seed control the food. Today large corporations own most of the world's seeds, putting a stranglehold on the global food supply. You vote with your dollars when you buy seeds. If one of the reasons you want to grow your own food is to empower yourself by rejecting the huge industrial-agriculture complex, then buy seeds that are heirloom, open pollinated, rare, or endangered. Growing and saving heritage seeds helps to preserve genetic diversity while producing tasty home-grown vegetables. Buy seeds from local businesses or from companies online that specialize in heritage seeds from your region. You may discover a variety that is specially suited to your area that has been grown and saved for over a century!

Get to know your vegetables

It's okay to grow the same varieties of vegetables year after year. Each variety grows slightly differently. It takes time to really know the unique growing habits of each variety. The better you know a vegetable variety, the better you will be able to grow it. Getting to know a plant takes at least two years; the first year is all about observation of the plant and its habits. The second year you will start to recognize patterns. After you have found a few varieties that taste great and grow well for you, try adding one or two new plants each season.

Planting Seeds
Sowing seeds outdoors

There are a lot of great advantages to planting your seeds directly in the garden rather than starting them indoors. Early spring plantings may receive enough rain and won't need as much additional water. A young plant may also receive more light than they would indoors, yielding a healthier and sturdier plant.

Temperature

Success with seeds sown directly in the garden depends on three key factors: Soil temperature, seed depth, and moisture. To determine your soil temperature, you can purchase an inexpensive soil thermometer at most garden stores. You can estimate the soil temperature based on the average between the week's high and low temperatures.

Depth

The seed tells you how deep it wants to be planted. Bury to a depth of two to three times the thickness or diameter of the seed. Big seeds, such as peas, beans, and corn, are planted in holes or furrows. Small seeds can be sown on top of the soil and then covered lightly with compost or soil. Plant seed to the proper depth, cover it loosely with soil:

COOL

SEASON CROPS

REQUIREMENTS

Vegetable varieties that grow when air temperatures are roughly 40–60 degrees Fahrenheit. Soil temperature should be 45–55 degrees for optimal germination.

CROPS

Spinach, Swiss chard, cabbage tribe, peas, cilantro, lettuce, beets

WARM

SEASON OR HEAT CROPS

REQUIREMENTS

Vegetable varieties that require temperatures above 60 degrees to grow. Soil temperature should be 55–65 degrees Fahrenheit or higher for germination.

CROPS

Tomatoes, tomatillos, beans, squash, cucumbers, peppers, eggplant, corn

TIP FOR PLANTING VERY SMALL SEEDS

It is hard to sow very small seeds, such as lettuce or carrots, evenly across a garden bed. These tiny seeds stick together and germinate in clumps and clusters. For more even distribution, mix the seeds in a cup of sand. Scatter the sand evenly over your bed, then cover the sand with a thin layer of soil or compost (just enough to cover the sand). Pat the surface of the bed gently. Water as you would any seedbed.

pat gently so that the surfaces of the seed are touching soil and the air spaces are gone.

Moisture

For seeds to germinate, they need to be kept uniformly moist. Water your seedbed with a fine mist until it looks like pudding. Stop watering and let the soil absorb any moisture. Repeat. Keep beds consistently moist until seeds germinate. You may need to water seedbeds twice a day during warm weather. Laying a single layer of burlap over seedbeds in hot weather will keep soil moist and aid in germination. Just water the burlap, lifting to make sure that the soil is getting wet. Remove the burlap cover as soon as seeds germinate.

Spacing and thinning

Thinning seedlings is essential for producing decent crops. Thinning seedlings is one of the biggest challenges for all gardeners. It is hard to kill something that you have grown. Without thinning, you will end up with wimpy, spindly versions of real vegetables. These seedlings will be prone to disease and pest infestation and will yield little for the table. If you have trouble thinning lettuce year after year, try planting starts. Give each baby start 6 to 8 inches of space all around. When in doubt, allow a little more space between plants.

Plant spacing is important for abundant vegetable production and disease control. Allowing space for air to circulate around plants will help reduce the risk of fungal disease, such as blight or powdery mildew. Follow spacing suggestions on your seed packet and see what happens. You will learn through experience how much space different crops need. Even the smallest seed becomes a whole huge plant and needs space to grow. Sow your seeds farther apart and thin them so they have room to grow. Though it seems counterintuitive, the fewer plants in a garden bed, the bigger the yield. With more space to grow and less competition for water and nutrients, plants will produce larger leaves, roots, and fruit.

Weeds

A weed is any plant that is growing in a place where you don't want it. Getting to know your weeds is fascinating. Start by observing weed seedlings and again after they germinate and grow. Most gardens have a few weeds that appear predictably after you start irrigating the soil. Use a digital camera to take pictures of unfamiliar weeds as young seedlings and after they have made flowers and seeds. After the weeds have matured, use weed books to identify what you have.

Knowing what your weeds look like as they grow will help you know what can stay and what should go. A challenge of sowing seeds directly is that weeds germinate right along with what you intentionally plant. Put your seeds in rows so you can tell what you planted. Avoid seed blends, such as mesclun salad mix, until you know your weeds. Many weeds are edible, even desirable, since they protect the soil from compaction and make flowers for bugs. They do, however, compete for water, soil nutrients, and space. Keep your beds mostly weed-free for easy harvesting and bigger yields.

How to Grow Your Own Transplants

Starting a few seeds indoors is easy and fun. You will need a supplemental light source, containers, soil, and easy access to water. You'll also need a workspace where plants can grow for a couple of months, located in a convenient place for daily care and monitoring.

Light for growing transplants

For most indoor growing, you will need supplemental light. You may have great luck growing a few things in a sunny window. If seedlings become leggy, however, they need more light. Inexpensive 4-foot shop lights with cool fluorescent tubes, which can be purchased at most hardware stores, are a great source of supplemental light. They closely mimic the natural color spectrum of sunlight and use little electricity. Since fluorescent tubes are cool, you can hang the lights 1 to 2 inches above the tops of seedlings without burning the leaves and will prevent seedlings from becoming leggy.

Hang lights from chains above your seedlings to easily adjust the height as plants grow. If you have limited space indoors, lights can be placed in closets or under tables. Seedlings will need 12 hours of light and 12 hours of dark to grow properly. An inexpensive timer will allow you to control lighting periods.

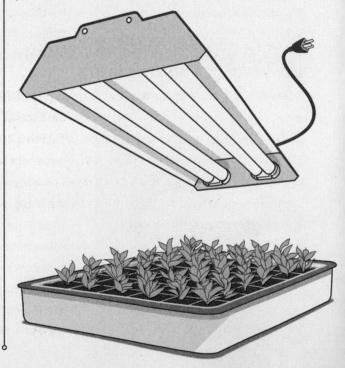

Watch your temperature

Seeds need consistent temperatures to germinate. Most vegetable seeds need a temperature of 50–75 degrees Fahrenheit in order to sprout. A heated home will offer enough heat for plants to grow—check your seed packet for the appropriate temperature range. Warm-season crops will appreciate 60–70 degrees Fahrenheit. You may need to keep cool-season plants in a garage or an unheated room in the house. They prefer to grow in temperatures between 50 and 60 degrees Fahrenheit.

Sterile soil a must

Use sterile starting mix, which is soil-free, or fresh potting soil to give new seedlings the best start.

Preparing containers

Vegetable starts can be grown in any 3- or 4-inch-square or round container that has a few drainage holes. If you are recycling old 4-inch plant pots (from last year's transplants), clean them and rinse them in a bleach solution (1 tablespoon bleach to 1 gallon of water) to kill disease organisms.

Liquid fertilizer: Nourishment for your growing plants

As seedlings grow, they quickly use up the nutrients in your planting medium. Starts may need more time to grow or they may need to wait for the weather to warm up before they can be transplanted. If your seedlings turn yellow, give them a quick burst of nutrients by watering them with liquid fertilizer (available in most garden centers) every two to three weeks.

Watering

Seedlings need consistent water for the first several weeks. It is helpful to have a water source close to your plants. Premoisten your planting mix before sowing and

Paper Pots

You don't need fancy pots to start seeds. You can make your own seed pots with a few common household items. You will need a strip of newspaper, some masking tape, potting soil, a water bottle or small glass, and seeds to plant.

- Wrap a strip of newspaper about 6 inches wide and 14 inches long around a water bottle or small glass so that a couple of inches extend over the bottom.

- Tape the seam on the side of the newspaper.

- Fold the bottom so that the newspaper overlaps and covers the bottom.

- Put another piece of tape on the bottom and remove water bottle.

- Fold down an inch of paper around the top to make the pot sturdier.

- Fill it with potting soil and plant your seeds.

- When your seedling is ready to transplant, remove the tape, open up the bottom and plant the whole thing—pot and all! Don't worry: The newspaper will decompose.

use a pump sprayer to lightly mist soil until seeds germinate.

Avoiding a mess

Soil and water can make a colossal mess. Protect carpet, wood, or other things you don't want to damage. Your propagation area will be used for two or three months in early spring. Washable surfaces are easy to keep clean and dry. Arrange pots on a tray to catch excess water. Keep the drip pan empty to regulate humidity and discourage fungus gnats.

Keep them close

Your seeds will need intense care as they germinate and grow. It is thrilling to watch seeds grow and to monitor them for moisture and problems. Place your propagation area where you can check on your babies daily.

Watch them grow

Continue to monitor for light exposure, moisture, temperature, and signs of disease as your seedlings grow. Seeds do not need any light until they have germinated and you see their heads poking through the soil. Once seedlings have emerged, they will need 12 hours of light and 12 hours of darkness.

Keep the soil evenly moist until seedlings emerge. After seedlings emerge, water them less. Let the soil surface dry out between waterings. Keep a close eye on your seedlings and use your finger to test for soil moisture.

When your seedlings have two pairs of leaves, they are ready to be transplanted.

Indoor seed-starting hints

Seedlings grown inside under lights can become leggy. These seedlings are weak and more susceptible to insects and disease. Keep lights close to your seedlings and provide 12 hours of light. After seedlings have gotten spindly, they are not good candidates for your garden. Resow seeds, keep light close, and monitor temperature. Cooler temperatures promote stockier, slower growth.

Humid indoor seed-starting areas can encourage fungus gnats and dampening-off disease. Keep things clean—wash pots in a bleach solution, provide good ventilation, and avoid overwatering. Fungus gnats are tiny insects that live on the surface of the soil. They are harmless but annoying, especially if you have them in your house. Use sterile starting mix, allow soil to dry out between waterings, and increase air circulation to keep them away.

If an otherwise healthy–looking plant develops a dark, shriveled ring around the stem at the soil level, then keels over and dies in a day's time, it is suffering from dampening-off disease. Dampening-off is a fungal disease that thrives in stagnant air, high humidity, and moisture. Fungus spores attack new seedling stems at the soil level, cut off the water and nutrient supply, and kill the plant. Affected plants rarely recover.

To prevent dampening-off, use clean containers and sterile soil mix. Thin seedlings to increase air circulation or use a small fan if ventilation is poor. Reduce humidity by letting the soil dry out a bit between waterings. Try chamomile spray to prevent dampening off. Use on soil, seedlings, and in any humid planting area. To make chamomile spray, pour 2 cups of boiling water over ¼ cup chamomile blossoms. Let the mixture steep, then cool and strain it. Chamomile spray will keep for one week in the refrigerator.

Troubleshooting Plants from Seeds

Symptom	Reason	Remedy
Low or no germination	Not enough water	Keep consistently moist while germinating.
	Old seed	Purchase fresh seed.
	Planted too deeply	Bury to a depth 2 to 3 times the diameter of the seed.
	Not warm enough	Outdoors: Resow after soil has warmed up; indoors: Add supplemental heat.
	Too hot	Outdoors: Some plants won't germinate when it is too hot; try a cooler microclimate or use a shade cloth (found at garden centers) to block the sun.
Seedlings are spindly and leggy	Not enough light; in the dark too long after germinating	Outdoors: Make sure the area gets at least 6 hours of sunlight. Indoors: Keep light close to seeds and starts; cool florescent lights should hang 2 inches above seedlings; put lights on a timer on a 12-hour cycle.
	Too crowded	Thin or resow, giving each seed more space to grow out rather than stretch up.
	Soil is too warm	Outdoors: Use a shade cloth or water in the evening to help cool soil. Indoors: Move to a cooler place, run a small fan to cool things down.
Seedlings shrivel and keel over	Dampening-off disease	Thin seedlings and increase air circulation; don't over water; use a small fan to aid in ventilation; use clean pots, flats, and sterile starting mix.
Leaves turn yellow	Start has used up the nutrients in the starting mix	Water with liquid fertilizer to provide instant nutrients.

HARDENING-OFF BEFORE *Transplanting*

Hardening-off is the process of easing your vegetable starts from indoor to outdoor growing conditions. The difference between your house at 67 degrees Fahrenheit and a spring night at 40 degrees is abrupt and can kill your seedlings.

Put your starts outside for increasing periods of time over the course of a week or two. Gradually increase your plants' exposure to full, direct sunlight, wind, and outside temperatures. After a week or two, leave plant starts outside overnight in a protected area under a large cardboard box or floating row cover. After two weeks, your starts should be ready to go in the ground. Without this gradual hardening-off process, your plants may not survive being transplanted.

You can also use a cold frame or cloche to help with the transition outside. These mini-greenhouses will regulate temperature and protect your plants from winds and rain. You can vent the cloche during the day for increasing periods of time throughout the hardening-off process.

Transplanting

You may transplant your seedlings into a larger container or directly into the garden. Transplant a seedling into the ground after it has developed at least two sets of true leaves or when the roots start to grow out the bottom of the pot. Use potting soil for transplanting seedlings into larger containers. If you are transplanting into a larger pot, increase your pot size by one size, or 1 to 2 inches larger on all sides.

Transplant your starts into a garden bed that has been amended with compost and fertilizer. If you are spot-planting your starts, dig a hole a little bigger and deeper than needed. Add some compost and a little fertilizer to the hole and mix thoroughly. Then plant. Topdress your transplants with half-an-inch of compost.

HOW TO PLANT SEEDS IN POTS

1. Start with clean pots and fresh planting mix.
2. Lightly moisten soil with water.
3. Fill pot to the brim and tap on the bottom to settle the planting medium.
4. Plant seeds in a hole or scatter them on top of the soil and cover them lightly with more soil. Plant seeds to a depth of 2 to 3 times their diameter.
5. Plant 1 seed in a small pot (½ cup soil or less) or 2 to 5 seeds in larger pots (1 cup or more of soil).
6. Pat the soil gently.
7. Water with a misting head or pump sprayer.
8. Label each pot.
9. Put seeds under lights after they have germinated.

How to Transplant Starts into the Ground

Use a trowel or hori hori to dig a deep well in the soil that is as big as your pot. If the soil is wet enough, it will stick easily to the sides and not fall into the hole. If the soil is too dry, water the area with a sprinkler and wait several hours before planting.

1. Gently put your hand across the top of the pot with the stems between your fingers. Flip the pot over and gently squeeze or tap the pot and remove the plant and soil.
2. The root ball will be a coil of roots. Loosen them.
3. Hold the stem and dangle the roots in the hole. While suspending the start in the hole, backfill with the soil. Starts should be planted to the same depth that they are in the pot. Tomatoes and tomatillos can be buried deeper (see chapter 7). With your fingers, tamp the soil gently but firmly around the stem.
4. Level the soil around your transplant so that water will not run off. Add more soil to create a hill, so that water will pool and slowly be absorbed around the root ball.
5. Water well.
6. Label each plant.
7. Topdress with compost.

Transplanting seedlings

Pricking out is the term for gently moving very tiny seedlings into a larger container. After seedlings have developed one pair of true leaves, they can be transferred to a bigger container to grow some more before you plant them in the ground. Fill your pots three-quarters full and use a dull knife to dig a deep well in the potting soil. Gently loosen and scoop the seedling from the planting medium. Lift the seedling by holding the leaves, not the stem. The stem of a small seedling is like its throat; holding it by the stem will kill it. Dangle the roots of the seedling in your pot and gently backfill with soil. Tamp the soil down and water. Label each pot. Transplant to the garden later.

Seed Saving

Any open-pollinated or heirloom seed can be grown and saved to plant again next season. Seed saving is a tradition that city farmers can also practice. Growing your own seeds saves money, closes the circle, preserves cultural diversity, and conserves rare vegetable varieties. Legumes, brassicas, and lettuce are great choices for first-time seed savers (tomatoes, squash, and peppers are a little more complicated to save as seeds).

A plant's purpose is to further its species by creating seeds. When a vegetable bolts and makes flowers, it is actually making seeds. Seeds are produced from the flower and are often found in a fruit or pod. Seeds are the plant's last dying chance to carry on its species.

Growing plants for seed takes time. The seeds must stay on the plant until they are ready; if you collect the seeds too early they won't be viable. The collard greens I saved this season took more than three months from flowering until the seed pods were dry and ready to save. Seeds are ready to save when they are brown and dry and come away from the mother plant easily. Withhold water from plants after seeds have formed to promote ripening.

Legumes

Save some legume seed by letting a few bean or pea pods dry out on your vines. The seed is ready when the pod is brown and the seed inside rattles. Collect pods and let them dry completely in a brown paper bag away from heat and sunlight. Crack open the dry pods to collect the dried seeds.

THE STORY OF SEEDS

A plant's main purpose is to further its species and to colonize. Most vegetables do this by creating seeds. Seeds enable plants to multiply and help plants move from place to place so that new seedlings have space to grow. Seeds move in many different ways—some are designed to just fall apart and drop around the base of the plant. Some seeds travel by wind, some have burrs that stick to animals or people, some are consumed and then go through the bodies of birds or other animals, and others explode and fly far from the mother plant.

This is how we might explain it to kids: "Now close your eyes and imagine that you are a little seed—a tiny little baby dandelion seed called an embryo. You are still asleep. Your mother is a big tall sturdy dandelion. She is going to send you on a really long journey. She can't come with you because she is rooted in the soil. Because she's a good mother, though, she gives you everything you need to make your journey and find a new place to grow—a raincoat, a lunchbox, and a hat.

"First, she wraps you up in a strong seed coat, which is bumpy and oblong. It is tight and warm, like a raincoat, to protect you while you are traveling.

"Your mother knows that you will be hungry when you finish your journey, so she also packs you a lunch. This lunch is full of nutrients and protein so you have enough energy to make your own leaves and start to make your own food from the sun. This lunchbox is the cotyledon.

"You and your cotyledon are all wrapped up in your seed coat and, because you are a dandelion, your mother will give you something else. She will give you a big fluffy feather hat so that the wind can blow you far away. There you will have space to nestle into the soil and grow a big strong taproot. So you can grow up nice and strong and make seeds of your own."

Brassicas

Members of the cabbage tribe are cool-season crops that bolt easily in hot weather. The yellow- or cream-colored flowers are delicious in salads. Seed pods are long and thin. They are split down the center by a membrane and make a neat line of seeds down either side of this separator. When pods are brown and dry, each side splits and springs away from the center membrane, shooting seeds around and away from the mother plant. Collect seedpods when they start to turn brown or purple and when the seeds inside are brown. Let them dry completely in a brown paper bag away from heat and sunlight. Crack open the dry pods to save the seeds.

Lettuce

Lettuce flowers are like mini-dandelions and turn into dainty puffballs when the seeds are

Profile: SEED SAVERS EXCHANGE

Everyone used to save vegetable seeds, because if you didn't save seeds from your harvest, you would have nothing to plant the next season. Over time, seed companies did the saving for us. Often, their requirements for seeds were not taste, but a long shelf life at the grocery store. Old varieties, full of flavor and history, were in danger of being lost forever.

In 1975, Diane Ott Whealy and Kent Whealy started Seed Savers Exchange (SSE), based in Decorah, Iowa. They opened their own seed stash to other gardeners and asked gardeners to share theirs. Their mission: To save the world's diverse, but endangered, garden heritage for future generations. The more varieties of seeds we maintain, the less likely it is that a pest or disease could decimate or even totally destroy one kind of vegetable. SSE now protects more than 25,000 different kinds of plants from extinction.

Many of the seeds are heirloom, but some are modern selections. The most important thing is that they are open-pollinated, which means that gardeners can save seeds from their own crop to plant next year or to share with others.

Seed Savers Exchange sells seeds, shares information on how to save seeds, maintains an orchard of old fruit varieties on its 890-acre farm, and supports the preservation of an ancient breed of cattle.

LEARN MORE:
seedsavers.org.

mature. The long, slender seeds are attached to fluffy parachutes to fly away with the breeze. Collect seeds when you can pull the puffball easily from the flower stalk with a few seeds still attached. Cut flower stalks and let them dry completely in a brown paper bag away from heat and sunlight. Shake the dry stems to collect the tiny seeds.

How to Get Started

Growing vegetables from seeds or starts is exciting. Delicious food sprouts from each tiny seed or transplant. Be sure to pick things that grow easily in your climate and that are best suited for growing in raised beds or containers. Start your own mini-nursery and raise plant starts for family and friends.

Save rare and endangered seeds to preserve our botanical history.

SOIL FERTILITY

 OIL FERTILITY IS THE LEVEL of nutrients in the soil for optimal plant growth. One way to figure out how much fertilizer and what kinds of nutrients to add to your soil is to send a soil sample to a soil-testing lab. Tests performed by these labs determine the amount of nutrients and organic matter in your soil. Based on these tests, soil-testing labs provide recommendations for how much fertilizer is needed (if any) for healthy plant growth. Testing for new gardens provides a baseline for comparison as you work to build healthy soil. Most basic tests give you information about your soil's pH, organic matter, nutrients (NPK—nitrogen, phosphorous, and potassium—the big three), and lead. In addition to testing soil for nutrient levels, you may want to test for arsenic, pesticides, oil, or gasoline if you suspect soil contamination.

Kits for testing pH, nitrogen, phosphorous, and potassium can all be purchased at garden centers. Although it is fun to collect soil samples and pour powder into test tubes, the results from these tests can be hard to interpret and are often inconclusive. For accurate results and custom recommendations, use a soil-testing laboratory in your area.

TAKING A
Soil Sample

Test your soil fertility at the same time each year, ideally in the fall. Your soil-testing lab will have instructions about how to take and send in your sample. For most labs, you will collect 10 to 12 samples from the area you want to test. To collect a sample, use a clean spade or trowel and take a slice of soil 6 to 8 inches deep—include the soil from the surface to the bottom of your hole. Mix the samples in a bucket and then spread the soil out on newspaper or flattened cardboard to air-dry. Mix the soil again and package the sample as instructed by the testing lab. Avoid testing very wet soils.

Where to Send a Soil Sample

The University of Massachusetts Soil Testing Lab offers an inexpensive basic soil test (currently $9) for pH, nutrients, lead, and other heavy metals, with excellent recommendations for home gardeners about adjusting pH and nutrient levels. Download soil sampling instructions and the order form from the university's Web site: http://www.umass.edu/soiltest/order.htm. Test results are usually available in two to three weeks.

To find soil-testing labs in your area, Type "soil test lab" and your state into your favorite search engine , or contact your local cooperative extension service for regional sources.

Soil pH

The concentration of hydrogen ions in a solution is called pH; this reflects the level of acidity or alkalinity in the solution. Acidity versus alkalinity is measured on a scale of 1 to 14, with 7 being neutral. Lower pH is more acidic and higher pH is more alkaline. Maintaining proper pH is important in promoting soil fertility and health. The level of pH affects the availability of nutrients to plants. Vegetables need a neutral, slightly acidic pH to grow well. Most water-soluble nutrients are available to plants in a soil with a pH of 6 to 7.5. Acidic soils can be made less acidic with the addition of lime; the pH of alkaline soils can be lowered (made more acidic) with the addition of organic matter or sulfur.

Easy pH Testing

While most store-bought tests are unreliable, you can get a broad pH reading using litmus paper. Since most vegetables and herbs that grow on city farms like a soil pH of between 6 and 7.5, using litmus paper will determine if your soil is in the optimal range. It's easy. Collect a soil sample (about 1 cup) and make a slurry with distilled water. Dip in the litmus test strip and match the color with the chart provided. If your pH is close to 7, you are ready to grow most vegetables.

Major Plant Nutrients

There are three major nutrients all plants need to thrive: nitrogen, phosphorous, and potassium (abbreviated as NPK, using their chemical monikers). Other elements, such as calcium, magnesium, and sulfur, are considered micronutrients because plants use them in much lower amounts. All these nutrients are present in healthy soil. Adding organic fertilizer based on test results will ensure that nutrient levels are adequate for vegetable crops.

NITROGEN produces vigorous growth and promotes lush green leaves. Applying nitrogen early in the season or after transplanting will give plants a boost to promote their

pH Requirements for Edibles

4.5-5

Acid-loving fruits such as blueberries, cranberries

5.5-6

Tomatoes, potatoes, raspberries, strawberries, rye

6-7.5

Most vegetable crops except tomatoes and potatoes

7.5-8

Too alkaline for most crops

growth. Too much nitrogen can burn plants, making the leaves turn yellow: Wait 3 weeks after adding fresh organic matter or manure so materials can decompose. Organic sources of nitrogen include compost, legume-based cover crops, alfalfa meal, fish-bone meal, blood meal, and livestock manure.

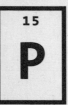

PHOSPHOROUS promotes the growth of fine root hairs, vigorous blooms, and healthy plant cells. Apply phosphorous to encourage abundant blooms and fruit. Sources of phosphorous include bone meal, fish-bone meal, and colloidal rock phosphate. Unlike nitrogen, phosphorous is released slowly and will still be available to plants for 2 to 3 years after it's added to the soil.

POTASSIUM, or potash, assists in the absorption of nitrogen, calcium, and trace minerals. It supports the overall growth and health of the plant and increases disease resistance. Potassium helps create strong roots, stems, flowers, and fruit. It improves the stamina and quality of all plants and is especially important for root crops, such as carrots, potatoes, radishes, and beets. Sources of potassium include wood ashes, greensand, granite dust, and kelp meal.

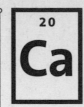

CALCIUM is essential for nitrogen absorbtion and protein synthesis in plants. Calcium builds sturdy cell walls. Lime is an excellent source of calcium. Lime helps particles stick together, increasing the pore space in soil and loosening and softening the soil. Soil testing will tell you how much lime to add to keep your pH in balance. Sources of calcium are dolomite and agricultural lime, wood ashes, bone meal, and oyster shells.

MAGNESIUM is an essential part of the chlorophyll molecule and is necessary for phosphorous metabolism. Sources of magnesium include dolomite lime, rock phosphate, and livestock manure.

SULFUR is an essential component of protein and fats. Sulfur deficiency is rarely a problem in soil, especially in soil where adequate organic-matter levels are maintained. Organic sulfur sources are organic matter and oak-leaf compost.

Trace elements are like vitamins for plants. Trace minerals, such as zinc, sulfur, manganese, molybdenum, and boron are essential for plant growth, but are needed only in small amounts. They are present in well-rotted compost and livestock manure.

Fertilizing

Nitrogen, phosphorous, and potassium (NPK) are present in the plants and other organic matter that goes into your compost. These nutrients are slowly released and made available to the plants with the help of some mighty tiny microbes (see chapter 3). For most plants, using compost and mulch will provide all the NPK and micronutrients required for healthy growth.

Compost alone won't meet the demands of intensive vegetable production, however. Annual vegetables have very high nutrient needs in order to produce large fruit, leaves, and roots. The nutrients in the vegetables grown in your garden come from the soil. Soil quickly becomes depleted with the demands of successive crops. As you take those nutrients from the soil in the food you harvest, they must be replaced by adding compost and organic fertilizer.

Fertilizer provides the nutrients plants need to grow. It can be rotted manure, compost, or a pre-mixed blend, purchased at a garden store. Organic fertilizers are derived from natural sources such as ground-up animals (fish-bone meal, blood meal), ground-up plants (alfalfa meal, kelp meal), or ground-up rock or minerals (rock phosphate, greensand). Organic fertilizers contain micronutrients as well as the big three—nitrogen, phosphorous, and potassium. Using organic fertilizers promotes soil building and increases the activity of soil organisms. Nutrients are released slowly, so they are available to plants for a long time, and you may not have to add fertilizer every season.

The numbers on the fertilizer box refer to the percentage of major nutrients in the mix. N refers to the amount of nitrogen, P to the amount of phosphorous, and K to the amount of potassium. For most vegetable crops, use a balanced fertilizer with nearly equal amounts of nitrogen, phosphorous, and potassium (3-2-2). If you are growing leafy greens, such as lettuce, spinach, and salad greens, look for a formula that has higher nitrogen and lower phosphorous and potassium (5-1-1). If you're growing fruiting crops, such as tomatoes or squash, buy a blend that has lower nitrogen and more phosphorous and potassium (5-7-3). If you aren't sure which fertilizer to choose, purchase a multipurpose organic vegetable fertilizer and use it for everything!

Purchase certified organic fertilizer to ensure that all the fertilizer sources were grown, cultivated, and processed without the use of chemicals, and follow strict standards set forth by the certifying agency (such as Oregon Tilth, OMRI, or USDA). Look at the ingredients; they should all be things that come from plants, animals, or minerals. Here are some common ingredients found in organic fertilizer. Alfalfa meal, kelp meal, crab meal, bone meal, feather meal, fish-bone meal, lime, mined gypsum, granite grit, and rock phosphate.

Symptoms of Nutrient Deficiencies in Your Soil

A soil test is the most accurate way to determine which nutrients are present in your soil, but sometimes you can diagnose deficiencies simply by looking at your plants.

MISSING NUTRIENT	SYMPTOM
Nitrogen	*Older leaves and veins turn yellow from the tips or the plant is stunted and pale green.*
Phosphorous	*Plants are stunted and have very dark purple-green hue. New leaves are pale with yellow edges; poor flowering and fruit development.*
Potassium	*Shortened (unnatural) distance between leaf nodes on the stem. Leaf tips turn yellow and appear scorched; weak stems; susceptible to disease.*
Calcium	*Poorly formed flower buds, leaf curl, poor root formation, and blossom-end rot in tomatoes and peppers. Plants lack resistance to fungal diseases and seedlings are affected by dampening-off.*
Magnesium	*Older leaves turn yellow, and whitish stripes appear on the leaves between the veins. Veins are often green with bronze spots on leaves.*
Sulfur	*Leaves are pale green and may turn yellowish; only a problem in new gardens with low organic matter or in gardens that are overwatered, since sulfur leaches easily from the soil.*

There are 3 types of organic fertilizer: dry or granular, liquid, and manure.

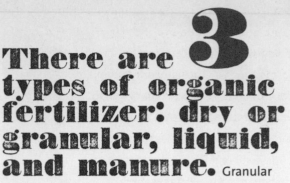

Granular can be loose meal or formed into tiny pellets. Liquid is a bottled concentrate that is diluted with water. Manure is the world's oldest source of slow-release plant nutrients. Granular and manure are slow-release fertilizers, while liquid is immediately available to plants.

Granular or dry fertilizer

This type of fertilizer releases slowly over the entire growing season, so it has to be applied only once. Dry fertilizers are not available to plants as soon as you add them to your soil. They first need to be broken down by soil organisms and converted to a form that plants can use. Soil microbes are dormant at temperatures below 45 degrees Fahrenheit. Nutrients in granular fertilizer will become available after the soil temperature has warmed up and microbes are active—late spring through early fall. Granular fertilizer can leach from the soil and pollute streams or lakes, so apply it at rates lower than those recommended on the package.

Sprinkle granular fertilizer over the surface of the soil and use a digging fork to mix it into the top 3 to 6 inches of soil. For spot-planting, mix thoroughly in the bottom of the hole with a couple of inches of compost. Another option is sidedressing, which means sprinkling a little fertilizer on the soil between plants and then lightly scratching it into the soil with your fingers or a forked-hand cultivator. Dry or granular fertilizers can be added in the spring just before planting or during the growing season. Add fertilizer to the entire bed when preparing your soil, when transplanting, or as a sidedressing around plants.

Manure

Manure is the oldest form of fertilizer. When people kept farm animals, they never needed to buy fertilizer or worry about soil fertility. Plant waste was given to animals and animal waste was recycled into the garden. If you are raising animals on your city farm, your garden will benefit from this valuable resource.

Fresh manure has too much nitrogen to use on plants, meaning it will burn plants if it's applied directly. Options for using fresh manure include adding it in the spring, a month before planting, or in the fall, as you put your garden to bed for the season. This adds nutrients when you most need them. Manure can also be composted with yard waste and added to garden beds a few weeks before planting. Bagged manure products available at garden stores or nurseries have been dried and shouldn't burn plants. Apply dried manure sparingly because the nutrients are more concentrated with the water removed.

Liquid fertilizer

Liquid fertilizer is water-soluble, immediately available to plants, and does not require soil organisms to break it down. Use liquid fertilizers in early spring when soil temperatures are too cold for much microbial activity. Apply it again in the summer as supplemental fertilizer to increase production of tomatoes and squash. Liquid fertilizer is the best choice for container plantings as well, since potting soil contains few microbes.

Dilute 1 to 2 tablespoons of liquid fertilizer with water in a watering can. Pour the amended water on the soil around your plants. Apply the liquid fertilizer every six to eight weeks for vegetables in garden beds and every two weeks for those in containers. Liquid fertilizer can be applied every few weeks throughout the growing season.

Crop Rotation

For the city farmer just starting out, there is a lot to factor in when planning and planting a vegetable garden. Adding one more technique may seem overwhelming, but crop rotation is one of the most important practices for organic food gardening. Crop rotation is the technique of changing crops planted in the same garden bed from one season to the next. Crop rotation increases the variety of

Plant Families and the Pests That Target Them

Nightshade family
(Solanaceae): Tomatoes, potatoes, peppers, and eggplant are members of this family. Flea beetles and aphids are pests associated with this group. Fungal diseases, such as late blight, also attack these plants.

Cabbage family
(Brassicaceae): Also referred to as "brassicas," this family includes broccoli, kale, cabbage, mustard greens, arugula, radishes, and turnips. Pests and diseases that attack this family include the cabbage looper moth, the imported cabbageworm, cutworms, club root, flea beetles, and aphids.

Onion family *(Alliaceae): Leeks, scallions, garlic, and shallots all belong to the onion family. Alliums are especially susceptible to fungal diseases, such as basal-root rot and rust.*
Source: COG

plants grown in your garden, encourages biological diversity, and maintains healthy soil.

There are two reasons to rotate crops— (1) for disease and pest prevention and (2) to promote soil fertility.

For a small city farm, rotating crops may be challenging. There are favorite crops that we like to grow year after year and, with limited space, crop rotation is often ignored by urban gardeners. But this will deplete the soil and invite pests. One extreme example of the problem of planting the same crops repeatedly is the 19th-century Irish potato famine, which was caused by growing the same variety of potato in the same soil year after year. Over time the potatoes depleted the soil of potassium and other nutrients. This led to a disastrous soil-borne fungal disease, which decimated potato production in Ireland.

Family Rotation

If crop rotation is new to you, start out by changing the location of crops that belong to the same family from one year to the next. For example, plant all your lettuce family together, all your beet relatives together, and so on. Then the following year plant them in another spot in the garden. This will help you become familiar with the unique characteristics of the different plant families.

By rotating crops and planting plant families in a different spot each year, you will prevent diseases and pests from becoming a problem in your garden. Growing crops from one plant family in the same spot year after year promotes the growth of disease organisms and pests specific to that family.

For optimal pest and disease prevention, wait three to seven rotations before placing a plant family back in the same place where you began. Maintain soil fertility with modest compost and fertilizer application and yearly soil testing to monitor nutrient levels.

Crop Rotation by Family

BEET FAMILY (Chenopodiaceae)	MUSTARD FAMILY (Brassicaceae)	PEA FAMILY (Fabaceae)

beets
lamb's quarters
orach
quinoa
spinach
Swiss chard
	arugula
	Asian greens:
	  Bok choy, gai lan,
	  etc.
	broccoli
	brussels sprouts
	cabbage
	collard greens
	kale
	kohlrabi
	mustard greens
	radishes
	turnips
		beans: Snap, dry
		  field peas
		crimson clover
		fava beans
		peas: Snow, snap,
		  and shelling
		vetch

CARROT FAMILY (Apiaceae)	NIGHTSHADE FAMILY (Solanaceae)	SQUASH FAMILY (Cucurbitaceae)
carrot chervil cilantro dill fennel lovage parsley parsnip	eggplant tomatillos tomatoes peppers potatoes	cucumber melon pumpkin squash
CORN FAMILY (Gramineae)	ONION FAMILY (Alliaceae)	SUNFLOWER FAMILY (Asteraceae)
corn rye teff wheat	chives garlic leeks onions shallots	burdock endive escarole Jerusalem artichokes lettuce radicchio salsify scorzonera

Source: COG

Crop Rotation for Disease Prevention

Make a list of the plants you will grow, and organize them according to family. Consider planting crops that belong to the same family together—that is, plant Asian greens, mustard greens, and arugula together for a spicy salad bed. Create a simple map of your garden beds and assign a number to each bed. Make a list of the bed numbers and record what you will plant. Adjust your plant list or rotation plans so that crops from the same family aren't planted later in the season in the same place. For example, if you grew mustard greens in Bed 1 in the spring, you should not follow with radishes or turnips in the summer because they are also members of the brassica family. Instead, follow the mustard greens with beets or carrots, which are root crops from other families. Again, allow three to seven rotations before placing a plant family back in the same place where you began. You may rotate using shorter intervals before replanting a family if there are no signs of disease or pests—increase the length of time between families if plants are affected by disease. Move families to new beds each year. Keep rotating.

Sample Two-Year Crop Rotation by Garden Bed

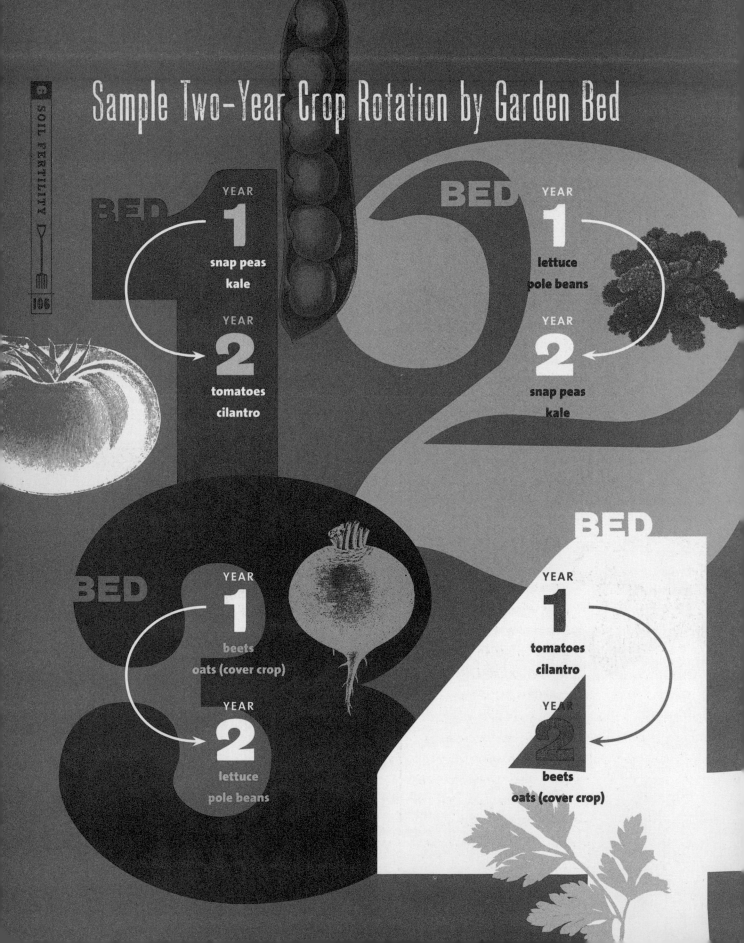

BED 1

YEAR 1
snap peas
kale

YEAR 2
tomatoes
cilantro

BED 2

YEAR 1
lettuce
pole beans

YEAR 2
snap peas
kale

BED 3

YEAR 1
beets
oats (cover crop)

YEAR 2
lettuce
pole beans

BED 4

YEAR 1
tomatoes
cilantro

YEAR 2
beets
oats (cover crop)

YEAR	BED 1	BED 2	BED 3	BED 4
1	snap peas	lettuce	tomatoes	beets
	kale	pole beans	cilantro	Cover crop: oats
2	tomatoes	collard greens	beets	lettuce
	spinach	summer squash	Cover crop: rye	cucumbers
3	carrots	beets	shelling peas	kale
	Cover crop: wheat	tomatoes	lettuce	runner beans
4	lettuce	snow peas	collard greens	Swiss chard
	bush beans	Cover crop: rye	cucumbers	tomatoes

What Are Heavy Feeders?

Some vegetables, known as heavy feeders, take up more nutrients from the soil and deplete the supply more quickly than others. Other vegetables are light feeders, meaning they use only a small amount of nutrients from the soil. Legumes and cereal grains improve the soil by fixing nitrogen and building tilth. As you create your crop rotation plan, try following heavy feeders with crops that improve the soil or "feed" more lightly. Think about it this way: Vegetables like broccoli and cauliflower are good for you to eat because they are high in vitamins and minerals. The vitamins and minerals in these vegetables are taken from the soil, which makes them heavy feeders.

HEAVY versus Light Feeders

HEAVY FEEDERS	LIGHT FEEDERS	PLANTS THAT IMPROVE SOIL
broccoli	beets	Snap, pole, broad, lima, and soy beans
cauliflower	carrots	
celery	collard greens	
corn	garlic	Cereal grains, such as wheat, oats, rye, and barley
cucumber	kale	
leeks	leeks	
potatoes	lettuce	
spinach	onions	All legumes, especially peas
squash	peppers	
tomatoes	radishes	
	Swiss chard	

Crop Rotation for Fertility

Creating a crop rotation plan to maintain soil fertility is a bit more challenging. As you get the hang of rotating crops, consider integrating the fertility rotation to maximize soil vitality and minimize the need for added fertilizer. Crops grown for leaves (such as lettuce) and those grown for fruit (such as tomatoes) require different soil nutrients. Rotating crops, based on what part you eat, reduces the need for added fertilizer by conserving nutrients in the soil. Continue rotating by families and you will prevent disease and maintain soil fertility!

The three major plant nutrients—nitrogen, phosphorous, and potassium (NPK)—play an important role in rotating crops for fertility. Each vegetable plant uses different amounts of NPK, depending on which part of the plant you eat. For example, lettuce and basil use lots of nitrogen to help produce lush, green leaves but use very little phosphorous or potassium. Root crops, like radishes, require more potassium, and fruiting plants like tomatoes need more phosphorous for proper flower and fruit development. Rotating crops in the order of leaf, root, flower, or fruit will conserve soil nutrients by growing plants that have different nutritional needs in a sequence. You will avoid depleting your soil of the same nutrient year after year and will reduce the need for added fertilizer.

Fertility Rotation: Leaf, Root, Flower, Fruit

Crops are classified as leaf (L), root (R), flower (FL), or fruit (FR) crops based on the part of the plant that is eaten.

LEAF *(L)*	ROOT *(R)*	FLOWER *(FL)*	FRUIT *(FR)*
basil	beets	annual flowers:	beans
bulbing fennel	carrots	Nasturtiums,	corn
cabbage	Jerusalem artichokes	calendula, zinnias,	cucumber
celery	potatoes	dianthus, bachelor's	eggplant
cilantro	radishes	buttons	melon
collard greens	rutabaga	artichoke	peas
dill	turnips	broccoli	peppers
garlic		buckwheat	pumpkin
leeks		(cover crop)	squash
lettuce		cauliflower	strawberries
mustard greens		crimson clover	tomatillos
oats (cover crop)		(cover crop)	tomatoes
onions		phacelia (cover crop)	any crop allowed
parsley			to go to seed
rye (cover crop)			
salad greens			
spinach			
Swiss chard			
winter wheat (cover crop)			

Source: COG

Tips on Maintaining Soil Fertility

Build and maintain soil fertility for a productive, problem-free city farm. Test your soil annually to ensure you have the proper pH and to avoid adding too much fertilizer. Rotate crops to keep insects and diseases at bay and to maintain soil fertility.

Sample Backyard Rotation for Fertility

YEAR	BED 1	BED 2	BED 3	BED 4
1	*Add fertilizer and compost* **FL—Phacelia (cover crop) R—Turnips**	*Add fertilizer and compost* **R—Potatoes FL—Crimson clover (cover crop)**	*Add fertilizer and compost* **FL—Buckwheat (cover crop) L—Cabbage**	*Add fertilizer and compost* **FR—Tomatoes L—Garlic**
2	**FR—Snap peas L—Rye (cover crop)** *Add fertilizer and compost* **FL—Buckwheat**	**FR—Peppers and eggplant L—Garlic**	**R—Potatoes**	**R—Beets FL—Fava beans (cover crop)**
3	**R—Carrots FR—Wheat (for seed)**	*Add fertilizer and compost* **R—Fall carrots FL—Field peas (cover crop)**	**FR—Fava beans (to eat)** *Add fertilizer and compost* **FL—Flowers for cutting L—Garlic**	*Add fertilizer and compost* **FR—Melons L—Lettuce**
4	**L—Kale FL—Field peas** *(no fertilizer needed for cover crop)*	**FR—Winter squash L—Oats**	**R—Fall beets FR—Amaranth (for seed) FL—Crimson clover (cover crop)** *(no fertilizer needed for cover crop)*	**R—Carrots FL—Crimson clover (cover crop)**

Source: COG

Profile: PERMACULTURE NOW!

Perma-what? The word may be unfamiliar, but the concept is easy to understand: Permaculture is the practice of sustainable living. Permaculture landscapes are designed so that little is needed from the outside; instead, by recycling all forms of energy, resources are continually available.

There is a great synergy to permaculture—the belief that the whole is greater than the sum of its parts. It encompasses everything from city planning to planting an apple tree in your backyard.

Permaculture is an old concept, but today's permaculture movement began in the 1970s in Australia and the organized idea spread from there. Permaculture design instructors are teaching courses all over the world, bringing these practices to urban communities.

One group of landscape designers, called Permaculture Now, share design practices and skills for building thriving and abundant communities in the Pacific Northwest, Central America, and Hawaii. This team of superstars collaborates extensively with a dynamic network of permaculturalists throughout the world.

Through hands-on methods, such as working on real projects at demonstration sites or on permaculture homesteads, urban folks can learn how to make the most out of their city lot. In community workshops, you can learn how to plant an edible hedge, grow edible mushrooms, design a legal gray-water filtration system or install a solar hot-water heater.

Interested in learning more about permaculture? There are activities taking place near you. Just enter "Permaculture Activities" and your state name in your favorite Web browser and join in the great reskilling!

LEARN MORE:
permaculture.org
AND permaculturenow.com

THE FRUITS & VEGETABLES OF YOUR LABORS

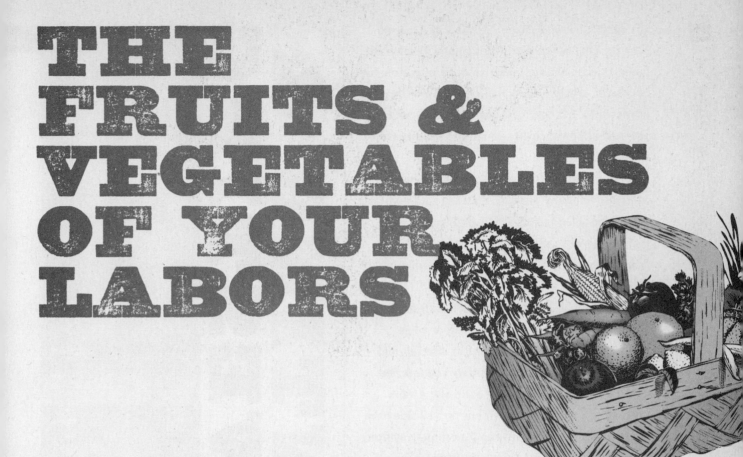

Vegetables

The vast majority of vegetable and fruit plants want just what we all want—a safe place to grow, with adequate water, sunlight, and nutrition. Doesn't that sound like a recipe for success? This chapter details the vegetables and fruits that respond well to that general care. In the unusual case where a plant needs more or less of something, it's also noted here.

The vegetable and fruit list is not exhaustive, because that would exhaust you. Instead, what you have here is a list of the best and easiest-to-grow top choices for your city farm. When you taste success—and you will—you will want to grow more and perhaps other varieties.

For information on frost dates (always important when you are starting from seeds) and a planting calendar, see pg. 112.

Last Frost Dates

Much of the success of your city farm depends on timing. Plant too early and seedlings get hit by a freeze or seeds don't germinate in cold, wet soil. Plant too late and the crop may not have enough time to grow to harvest size before short days and cool nights set in.

The first day of spring is not enough of an indicator, because climates across the country vary so much. But knowing your local last frost date of the year does help you plan for planting times to get the biggest harvest. Of course, the actual last frost date can change from year to year and (what with global climate change) surprises may be in store for your local weather. Your last frost date may be listed as March 31, but there are years when you won't see frost after March 1 or you might see a late one on April 15.

The listed last frost date for an area is an average over many years, so it's not a hard-and-fast day on the calendar. It's a little bit of a guessing game, but one that you'll get better at the longer you garden and get to know your area's unique climate.

Microclimates can affect your last frost date, too. Land that is depressed, with either trees, shrubs, or earth banks all around it, can trap cold air and damage plants that wouldn't be touched by frost if they had been at the top of the bank (instead of 3 feet lower). Gardens on a hillside have a great advantage, because cold air runs away from your garden down the hill as long as there are no obstacles.

The combination of knowledge of your own microclimate and your area's last frost date gives you more power over how your garden grows. To find the last frost date in your area, check the National Climatic Data Center—go to: cdo.ncdc. noaa.gov and search "freeze/frost data" for your region. Your state's Cooperative Extension Service will have frost date information or type "last frost date MY TOWN" into your favorite search engine.

PLANTING & HARVESTING

CROP	JA
arugula	
beans—bush	
beans—pole	
beets	
broccoli/cauliflower	
carrots	
corn	
cucumber	
eggplant	
kale/collards	
leeks	
lettuce	
onions	
peas—snow, snap, shelling	
peppers	
radishes	
spinach	
squash—summer	
squash—winter	
Swiss chard	
tomatoes	

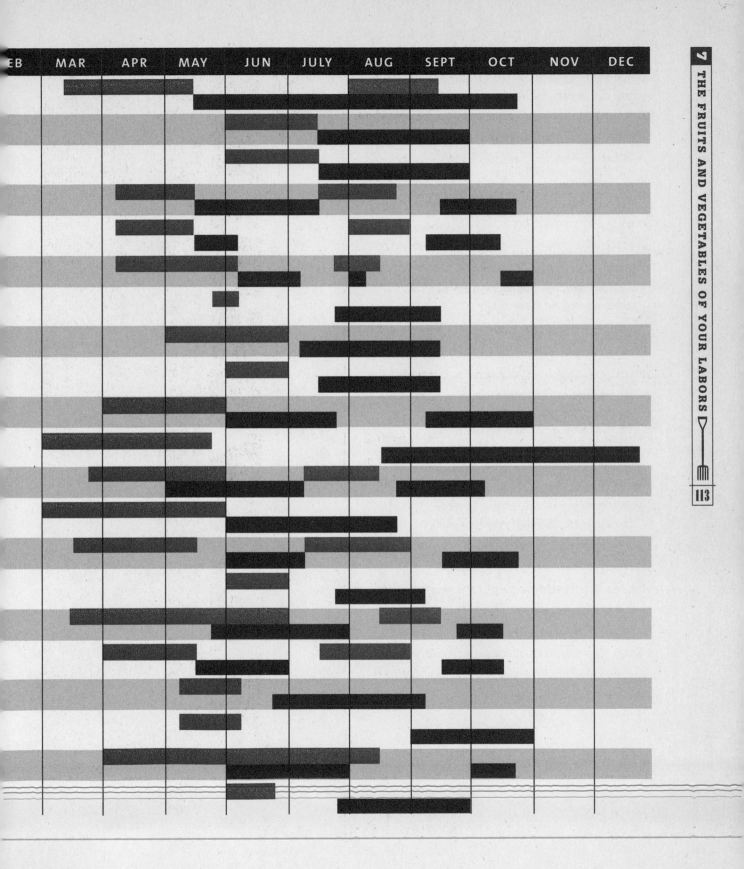

EB | MAR | APR | MAY | JUN | JULY | AUG | SEPT | OCT | NOV | DEC

7

THE FRUITS AND VEGETABLES OF YOUR LABORS

113

How to grow

- Plant arugula in early spring and in late summer for a fall harvest.

- Plant arugula in mid-April directly into the garden, sowing the seeds ¼-inch deep, 1 inch apart. Plants will emerge in 2–14 days.

- Thin the plants to 2 inches apart when they have two sets of true leaves.

- Plant successive crops every two weeks until mid-May when the weather gets too warm. Wait until late August, and start again.

How to harvest

- In addition to eating the thinnings, use scissors to cut baby arugula leaves off the plants just above ground level until the weather is warm and the leaves turn bitter.

- Arugula bolts quickly and the leaves become too hot and bitter to eat. Nibble on the sweet, smoky flowers as a quick snack when you are out in the garden or add to salads.

ARUGULA

AKA ROCKET (ERUCA SATIVA)

BEANS

(PHASEOLUS)

How to grow

- Bean types include bush (no trellising needed), French, pole, and runner beans. Beans can be grown for fresh eating (the whole pod or only beans) and drying (beans).

- Except for bush and French beans, create a string trellis or tepee for the beans to climb on (at least 6 feet high). The plants will snake up vertical lines and supports.

- Beans germinate best in warm soil—at least 60 degrees Fahrenheit. Use a soil thermometer or plant after your last frost date, figuring mid- to late May or even early June in cool-weather climates. If you plant beans too early, the seeds will rot. If that happens, just resow as weather gets warmer.

- Plant up to four beans per foot about 1½ inches deep, and space plants 6 to 8 inches apart. You won't have to thin them.

- Beans planted directly in the garden will appear in 8–16 days.

- Weed well but carefully, because beans are shallow-rooted.

- You can plant a crop of bush beans every couple of weeks; other beans once.

How to harvest

- Bush beans are ready to be harvested about 50 days after planting; generally, they produce one or two crops. Other beans are ready after about 70 days; they continue to produce if you continue to pick. Beans are ready to pick when you can feel the beans inside the pod, but the pod is still firm.

- Snap off beans at the thin stem between vine and bean.

- Eat the whole pod of bush and pole (also called "snap") beans. Shelling beans can be eaten fresh or cooked if dry.

- Pick beans for fresh eating when pods are firm and crisp with small seeds.

- Pick beans for drying when leaves start to yellow and the pods are lumpy. You can pull plants up early and dry them in the garage. Shell them by hand and save them.

- Bean blossoms make a sweet crunchy snack right off the plant.

Varieties to grow

SNAP BUSH (including French, or haricots verts): Dragon Tongue, Roma, Maxibel, Royal Burgundy, Slenderette

SNAP POLE: Kentucky Wonder, Blue Lake Pole, Musica, Romano, Cherokee Trail of Tears

RUNNER: Scarlet Emperor, Painted Lady

BUSH SHELLING: Black Coco, Tiger Eye, October, Jacob's Cattle

POLE SHELLING: True Red Cranberry, Mayflower, Wren's Egg

BEETS

(BETA VULGARIS)

How to grow

- Plant seeds in the garden beginning in early spring through late spring. Another crop can be planted in midsummer for fall harvest.

- Beet seeds like company so sow 1 or 2 seeds per inch, ½ inch deep and plan on thinning.

- When plants are 4 inches high, thin to 1–2 inches apart. Thin again and eat baby beets when tops are 8 inches tall, eventually thin to 4 inches apart.

How to harvest

- Beets grow as big as the space available, and you can harvest them anytime from about 40 to 45 days, depending on the variety.

- Eat beet greens early (they get tough as the roots mature), by cutting off a few leaves from each plant. Leave plants in the ground (and a few leaves on the plant) to continue growing.

- Eat both greens and baby beets when thinning.

- Eat beet greens in salads or stir-fry; roast beets for salads or a side dish. If you have a bumper crop, pickled beets are always crowd pleasers.

Varieties to grow

ROUND: Bull's Blood, Red Ace, Detriot, Dark Red, Early Wonder Tall

ODD-COLORED BEETS: Chioggia (red-and-white interior rings), Golden, Touchstone Gold, Yellow Mangel, Albina Vereduna (white)

LONG-ROOTED OR DANISH BEETS: Cylindra, Forono

Broccoli & Cauliflower

(BRASSICA OLERACEA)

How to grow

- Start seeds indoors by the end of March to transplant into the garden at the beginning of May.

- Sow seeds in individual pots and cover with ¼ inch of fine soil. Plants will emerge between 5 and 17 days.

- Harden-off seedlings off before planting outside by putting the pots outdoors during the day for a week.

- Set plants out 18–24 inches apart.

- Broccoli can be planted in succession, sow later crops directly in the garden.

- Plant sprouting broccoli which have edible leaves and produce small flowerets for several weeks.

How to harvest

- Harvest broccoli by cutting the stem off a few inches below the head; harvest cauliflower by cutting below the head (you may take a few leaves along with it).

- Harvest broccoli and cauliflower 50–100 days from planting in the garden; your plants should be ready by mid-June.

- Harvest cauliflower when you see the sections begin to separate.

- Harvest broccoli just before the tight buds begin to open and shoot up yellow flowers.

- Broccoli and cauliflower flowers can be eaten, too—either fresh or cooked

Varieties to grow

BROCCOLI: Southern Comet, Arcadia, Packman, Romanesco, DeCiccio

CAULIFLOWER: Snow Crown, Graffiti (it's purple), Cassius, Ravella, Early Snowball

SPROUTING BROCCOLI: Purple Peacock, Purple Sprouting, Broccoli Raab, Calabrese

CARROTS

(DAUCUS CAROTA)

How to grow

- Carrot seeds are tiny, but the eventual carrots can be long and need loose soil. Work soil only after it has dried out to a depth of 10–12 inches.

- Plant carrot seeds in the garden from mid-April through June (summer sowing for winter keepers).

- Space about 4 seeds per inch, plant seeds ½ inch deep and thin so there is 2 inches between plants. For easier sowing, mix seeds with sand (see chapter 5).

- Keep seeds moist until they germinate. Carrot seeds germinate slowly and may take 3–4 weeks to sprout in colder weather. Cover your soil with a layer of wet burlap to keep soil moist in summer heat.

- Thin the seedlings or your carrots won't grow. When leaves are 3 inches high, thin to a half-inch apart and when they are 6 inches high, thin plants to 1 inch apart. Mound soil around each plant to prevent the carrots from developing green tops, which taste bitter.

- Use a floating row cover to keep carrot rust flies from laying eggs at the base of the plants.

How to harvest

- Carrots are ready to harvest from 45 days for baby carrots to up to 100 days, depending on the variety.

- Check maturity by pulling a carrot to see what's there. You can pull and eat some as the rest continue to grow.

Varieties to grow

ROUND: Thumbelina, Parmex Baby Ball, Parisian, Romeo

SHORT (Nantes type): Baltimore, Scarlet Nantes, Nelson, Little Finger, Touchon, Oxheart, Bolero, Danvers

GOOD WINTER KEEPERS: Merida, Autumn King

CARROTS OF A DIFFERENT COLOR: Purple Haze, White Satin, Atomic Red, Amarillo

How to grow

- Corn will not germinate in soil cooler than 60 degrees, so wait until at least the end of May or plant indoors and transplant when the corn is 3–6 inches high.

- Plant corn in the garden 1 inch deep every 4 inches; the plants will emerge between 7 and 10 days. When the plants are 5 inches high, thin to one plant per foot.

- Corn is wind-pollinated, so you'll get the fullest ears by planting in blocks or rectangles—no less than 4 feet deep. Space plants 4-6 inches apart.

- Sweet corn and ornamental or popcorn must be planted 100 feet away from each other to reduce the chance of cross-pollination—you might want to choose to grow one kind or the other.

- Use a high-nitrogen fertilizer every 2 weeks until the tassels appear.

- Crows especially love the tender shoots and snap them off at soil level. Cover seedlings with a floating row cover until they are a foot tall.

How to harvest

- Corn will be ready to harvest 60–110 days after planting, depending on the variety, so you will be harvesting in August and September. One sign that it's ready is when the silks begin to turn brown and dry. You can also check an ear by peeling back the husk and poking a kernel, which should be full and release a milky liquid.

- Cool off the harvested corn quickly so that it retains its flavor. Eat it soon! We can't resist munching a sweet raw ear right there in the garden.

- If you grow popcorn or ornamental corn, husk the ears, and then spread the ears in a dry place for a couple of weeks to let them cure.

- Most types of corn grow only two ears per plant, so consider how much space you need versus how much you get to eat.

Varieties to grow

YELLOW SWEET AND SUPERSWEET: Sugar Buns, Bodacious, Ashworth, Golden Bantam

WHITE SWEET AND SUPERSWEET: Stowell's Evergreen, Silver Queen, Country Gentleman

BICOLORED SWEET: Rainbow Inca, Sugar Dots, Bloody Butcher

POPCORN: Calico, Dakota Black, Tom Thumb, Dwarf Strawberry

Corn
(ZEA MAYS)

CUCUMBERS

(CUCUMIS SATIVUS)

How to grow

- Cucumbers love warm weather, so don't sow seeds until the soil temperature is at least 60 degrees in late May through June.

- Build a trellis for the plants—it takes up less space and cucumbers stay off the ground.

- In the garden, plant seeds ½ inch deep, 4 seeds per hill, hills spaced 4 feet apart. After plants emerge, thin to 2 plants per group or about 12 inches apart.

- Start early indoors in paper pots and transplant outdoors after 3 weeks (be sure to harden-off the seedlings).

- Many cucumber varieties have separate male and female flowers on the same plant. The female flower has a small cucumber behind it; if it is not pollinated, it will wither and fall off.

- Water cucumbers regularly or they develop a bitter taste.

- Cucumbers can continue to produce until your first frost in the fall.

How to harvest

- Harvest cucumbers by cutting through the prickly stem just above the cucumber.

- Pick them young or when the variety chosen reaches its appropriate size, anywhere from 45–65 days from when the plants emerge.

- Keep picking and the plants will keep making cucumbers.

Varieties to grow

SLICING: Marketmore 76, Orient Express (an "English" or "burpless" type), Straight Eight, English Telegraph

PICKLING: Cool Breeze, Alibi, Boston Pickling Improved, Parisian Pickling, Russian

NOT YOUR MAMA'S CUCUMBER: Armenian, Boothby's Blond, Lemon, White Wonder, Painted Serpent

EGGPLANT

(SOLANUM MELONGENA)

How to grow

- Eggplants need a long, hot growing season, so if you live in a cool climate, plan on growing them under a cloche or hoop house.

- Start seeds indoors 8 weeks before your last frost date. Start seeds in pots on a warming tray that has heated soil to at least 75 degrees. Plants should emerge in 5–17 days.

- Harden-off transplants, and in June plant them outside using cloches or other heat-trapping covers.

- Plants should be 12 inches apart.

- Use tomato cages to support branches heavy with fruit.

How to harvest

- Harvest eggplants by using pruners or a sharp knife to cut through the stem just an inch or so above the eggplant.

- Harvest eggplants when the skin turns shiny, 55–75 days from planting, depending on the variety—August in most climates.

Varieties to grow

Fairy Tale, Black Beauty, Pingtung Long, Little Fingers, Little Spooky

Kale/ Collard Greens

(BRASSICA OLERACEA)

How to grow

- Start seeds indoors by the end of March, or direct-seed into the garden at the beginning of May.

- Plant seeds ½ inch deep, 1 inch apart. Plants will emerge in 5–17 days.

- Progressively thin your plants when the seedlings are about 3 inches tall so that they are eventually 12 to 24 inches apart.

- Collard greens can be planted until July.

- Start another round of kale in mid-July for winter.

How to harvest

- Eat what you thin—kale can be tossed in a salad and little collard plants can be lightly sautéed.

- Harvest kale and collard greens by cutting off the outer leaves at the base of the plant.

- Collard greens and kale can be harvested after 50 days, or pick individual leaves as needed.

- In regions without long, hard freezes, kale and collard greens can be left in the garden and harvested from there.

- If collard greens and kale are left in the garden, the following spring they will flower; their flowers taste like broccoli dipped in honey.

Varieties to grow

KALE: Lacinato, Nero di Toscana, Tuscan Kale, Winterbor, Redbor, Russian Red

COLLARD GREENS: Champion, Flash, Georgia, Green Glaze, Morris Heading

LEEKS

(ALLIUM PORRUM)

How to grow

- Leeks take a long time, but require very little care—and the reward is worth the wait.

- Start seeds indoors 10 weeks before your last frost date by covering a 4-inch pot with seeds, then sifting ¼ inch of fine compost over the top.

- Maintain an even moisture level; seedlings will emerge in 6–16 days, looking like green hair.

- Plant out 5 inches apart in the garden in a trench 8 inches deep.

- Gradually fill the trench so that much of the root end of the stem is covered; this will create a long, white stem.

- Leek starts may be available from your local nursery.

- Plant fall and winter leek varieties staggered a month or two apart for a nonstop harvest from late summer through early spring.

How to harvest

- Harvest anytime after the leek is ½ inch wide, which may be as early as July.

- Leeks keep in the garden well into fall and through the winter in mild climates.

Varieties to grow

King Richard, Lincoln, Lancelot, Falltime, Giant Winter, Giant Musselburg, St. Victor, Durabel

LETTUCE

(LACTUCA SATIVA)

How to grow

- Lettuce is a cool-weather crop that grows best when sown in successive crops in spring and again in late summer for fall harvest.

- Start lettuce indoors in March to set out in April. Sow directly into the garden in April.

- Sprinkle seeds onto the soil surface, and then cover lightly with soil, pressing down so that the seeds are in contact with the soil. Plants will emerge in 5–15 days.

- Lettuce seeds are tiny, and you will probably overplant—that's OK. Thin the seedlings to about 8 inches apart and enjoy the harvested baby greens.

- Plant more lettuce every two weeks until the beginning of June or when your hot weather begins. Lettuce seed won't germinate at temperatures above 75 degrees, so start planting your fall crop in mid-August or early September.

- Encourage fast, leafy-green growth for sweet flavor. This requires ample nitrogen and water. If your lettuce is bitter, it didn't get enough water and didn't grow fast enough. Water consistently and apply liquid fertilizer every couple of weeks for the best flavor.

How to harvest

- Lettuce will be ready to harvest from 35 to 55 days from planting, depending on the variety.

- Harvest loose-leaf varieties by snipping leaves at the base every few days and letting the plant continue to grow.

- Head lettuce, such as romaine and buttercrunch, can be harvested by cutting at the base, leaving the "stump." An application of liquid high-nitrogen fertilizer will encourage it to grow a few smaller heads from the stump.

Varieties to grow

LOOSE-LEAF: Red Sails, Black-Seeded Simpson, Merlot, Deer Tongue, Australian Yellow Leaf, Green or Red Oakleaf

ROMAINE: Devil's Tongue, Petite Rouge, Forellenschluss, Jericho

BUTTERHEAD: Buttercrunch, Tom Thumb, Merveille des Quatre Saisons, Little Gem

MIXES: Grow a gourmet salad, "mesclun," or other greens mix for a variety of tastes from one seed packet

ONIONS

(ALLIUM CEPA)

How to grow

■ Onions grow green tops until the summer solstice, when day length begins to shorten; from that point on, onions put energy into growing the bulb. The more green tops aboveground, the bigger the onion.

■ Onion plants emerge from seed in 6–16 days.

■ Plant onion seeds in the garden in mid-March ½ inch deep. Indoors sow up to 70 seeds in a 4-inch pot and then transplant into the garden, separating the hairlike green strands.

■ Thin to at least 5 inches between plants.

■ You can buy onion "sets," which are bundles of small plants that look dried up. Set them out as transplants and they will do fine.

How to harvest

■ Onions take 60–100 days to be ready to harvest, and will be ready in July or August.

■ Cut off flowers stems so that the plant's energy goes into making a bulb, not a flower.

■ Stop watering when the tops begin to die back and fall over. When this starts, you can promote the process by stepping on (and so knocking over) the tops that haven't fallen.

■ After a week, harvest the onions, letting them "cure"—that is, "paper" or form a dry outer shell—before storing.

Varieties to grow

Walla Walla Sweet, Cipollini Borrettana, White Sweet Spanish, Redwing (purple), New York Early, Red Long of Tropea

How to grow

■ Grow snow or sugar snap (for the whole pod) or shelling peas.

■ Install a string trellis. Peas climb by twining tendrils around horizontal supports.

■ Plant in the spring as soon as the soil can be worked.

■ Plant peas 2 to 4 inches apart and 1 inch deep.

■ Plant peas directly in the garden; they will appear in 6–14 days.

PEAS

(PISUM SATIVUM)

■ If your springs are wet and cold, sow peas in paper pots indoors and transplant them when they are 2–3 inches tall.

■ Plant peas in early spring; it's a good idea to plant a crop every 2 weeks until mid-May. If spring whizzes by before you can get your peas planted, sow another crop of snow peas in early August for an October harvest.

How to harvest

■ Snow peas are ready to eat 60 days after planting. Sugar snap and shelling peas will take 85 days before they are ready to harvest.

■ Pick snow peas when pods just begin to swell; harvest sugar snaps when the pods are filled out. These are eaten fresh or lightly cooked, such as in a stir-fry.

■ Pick shelling varieties when the pods are lumpy with peas.

■ Pea tendrils—the top 6 inches of the plant—can be pinched off and eaten raw in salads or used in stir-fry or lightly cooked in pasta dishes.

■ Pea blossoms are also edible—straight from the vine or tossed in a salad. The flowers of ornamental "sweet peas" are not edible.

Varieties to grow

SNOW: Oregon Giant, Oregon Sugar Pod II, Dwarf Grey Sugar, Mammoth Melting

SNAP: Cascadia, Super Sugar Snap, Sugar Daddy, Amish Snap

SHELLING: Alderman, Homesteader, Maestro, Canoe, Petit Pois, Green Arrow, Little Marvel (bush pea)

PEPPERS

(CAPSICUM)

How to grow

- Peppers need a long, hot growing season to ripen.
- Start seeds indoors 10 weeks before your last frost date in pots set on a warming tray that brings soil temperature up to at least 70 degrees. Plants will emerge in 8–25 days.
- Plant outside in the garden when the temperature has reached 60 degrees and overnight temperatures do not fall below 50 degrees. Plants should be 12-14 inches apart.
- Cloches or other heat-trapping covers will increase the ambient temperature.
- Remove protection when summer temperatures get high, because peppers will not set fruit if it's too hot.

How to harvest

- Harvest peppers by cutting through the stem just above the pepper.
- Peppers are ripe from 55–100 days after transplanting into the garden, depending on the variety, so your harvest will begin in July or August.
- Peppers are edible when green but will ripen to yellow, red, or orange, depending on the variety.

Varieties to grow

SWEET: Wonder Bell, Gypsy, Italian Sweet, Klari Baby, Jimmy Nardello's, Little Bells, Peacework

HOT: Hungarian Hot Wax, Thai Hot, Early Jalapeño, Cayenne, Habenero, Czech Black

RADISHES

(RAPHANUS SATIVUS)

How to grow

- Grow two crops of radishes—one in spring and one in fall.

- Radishes are ready all at once, so consider just how many you will eat at one time.

- In mid-April or the beginning of August, sow directly in the garden, placing seeds ½ inch deep and ½ inch apart.

- Plants emerge in 4–11 days. As soon as they emerge, thin to 1–2 inches apart.

- Radishes need lots of water or they will be pithy and too hot and spicy to eat.

How to harvest

- Radishes are ready to harvest at 20–30 days, so plan on pulling them about three weeks after they start growing.

- Radishes bolt (send up a flower stalk) quickly, sending up tall stalks covered with pale pink flowers that are great tossed in a salad. The plump seed pods are edible, too—sweet and radishy and crunchy. Pick seed pods before they get stringy and toss in salads or stir-fry.

Varieties to grow

Easter Egg II, Amethyst, Icicle Short Top (white and long), Daikon, French Breakfast, Long Black Spanish

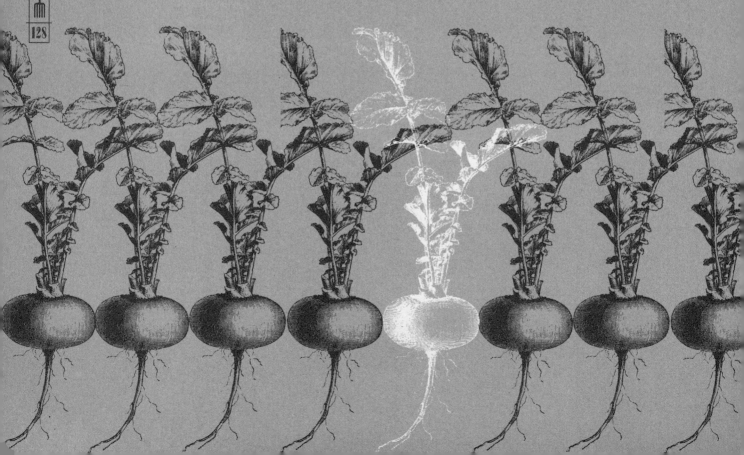

How to grow

- Spinach is a cool-weather crop that usually bolts in hot weather, so it's best to plant in early spring, and again in late summer for a fall harvest.

- Sow spinach seeds in early April or when soil temperature reaches 50 degrees or in August for fall harvesting. Plants started in September will overwinter and be the first available to harvest the following spring.

- Plant seeds ½ inch deep, 1 inch apart. Plants will emerge between 6 and 21 days.

- Thin to one plant every 3 inches.

- Succession-sow every week for an extended harvest.

How to harvest

- Harvest spinach by using scissors to snip through the stem at the base, just above soil level. Be careful not to damage the center of the plant.

- Harvest can begin when leaves are 3 inches long—just about three weeks after plants emerge.

- Otherwise, plants are ready anywhere from 35 to 50 days.

- If your plant bolts, try pinching off the stem and giving it some liquid fertilizer; it may grow back.

Varieties to grow

Tyee, Bloomsdale Savoy, Everlasting, America

SPINACH
(SPINACEA OLERACEA)

Summer (CUCURBITA PEPO)

SQU

How to grow

- Plant in mid- to late May outdoors, or when soil temperature has warmed to 60 degrees Fahrenheit.

- For a head start, sow seeds indoors in paper pots 3 weeks earlier than planting time.

- Plant seeds 1 inch deep, 6 inches apart, with 3–4 seeds per hill; space hills 4 feet apart. Plants will appear in 5–14 days. A squash "hill" is not a hill at all; it's just a term for a cluster of seeds planted together.

- Look at the flowers! All squash plants have separate male and female flowers on the same plant. A female flower has a little "squash" behind it. When the flower is pollinated, the squash grows; if the flower does not get pollinated, the squash withers and falls off. You can hand-pollinate your squash by using a feather or a small paintbrush to lift pollen from the male flowers and dust the females.

How to harvest

- Harvest squash by cutting through the stem just above the squash.

- Pick, pick, pick—you don't want to let a zucchini get away from you (although, you can roast the big ones). Harvest when summer squash is small, starting about 40 days after planting, and continue until fall.

- Summer squash can be used in any recipe that calls for eggplant. Try making a pan of Crookneck parmesan for that late summer potluck.

- Eat squash blossoms too: Pick male or female flowers (leave some to make fruit), and remove the center stamen or stigma. Squash blossoms are smooth, velvety and taste mildly squashy. They can be eaten fresh or stuffed, battered, and pan fried.

Varieties to grow

YELLOW ZUCCHINI: Gold Rush, Golden Zucchini

ZUCCHINI: Black Beauty, Bush Baby, Costata Romanesco, Cocozelle

PATTY PAN: Sunburst, Bennings Green Tint, Golden Scallop, White Patty Pan

UNUSUAL SHAPES: Trombocino or Trombetta (a climbing plant with squash shaped like a trombone), Yellow Crookneck

ROUND: Eight Ball, One Ball

Winter SQUASH

INCLUDING PUMPKINS (CUCURBITA)

How to grow

■ Follow directions for summer squash, except separate hills by 6 feet. Most winter squash are vining plants, and will shoot stems along the ground. Alternatively, they can be trained on a trellis.

■ Use a nylon sling to support heavy fruit hanging on a trellis.

How to harvest

■ Winter squash take from 80 to 100 days to grow to maturity; that makes them a fall crop.

■ Winter squash is ready to harvest when the squash has reached the appropriate color and developed a hard skin. Use your thumbnail to test squash skin. When the skin is firm and your thumbnail doesn't leave a mark, it is time to harvest.

■ After harvesting, cure the squash in the garden if your weather is dry (a light frost is OK) or store in a dry, dark place for 3 weeks. Squash should not be stored for a long time in a place where it's below 50 degrees.

■ Toast winter squash and pumpkin seeds for a high-protein snack. Clean the seeds, mix them with a little oil, salt, and a pinch of red pepper, and roast them.

Varieties to grow

WINTER SQUASH: Delicata, Sweet Dumpling, Buttercup (also called Kabocha), Butternut, Spaghetti, Hubbard, Acorn

PUMPKINS FOR EATING: Sugar Pie, Amish Pie, Yellow of Paris

PUMPKINS FOR CARVING: Howden, Magic Lanterns, Casper

ORNAMENTAL PUMPKINS: Jack Be Little, Rouge Vif d'Etampes, (Cinderella pumpkin), Dill's Atlantic Giant, Fairytale

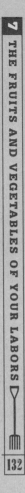

CHARD

(BETA VULGARIS)

How to grow

- Plant Swiss chard in the garden from mid-April into July.

- Sow seeds ½ inch deep, 3 inches apart.

- Plants will emerge in 5–17 days. Sow seed more densely and thin baby chard for salad mix.

- When plants are 3–4 inches tall, thin so that they are 12 inches apart.

How to harvest

- Chard is ready to harvest about 60 days from when it emerges, but you can harvest individual leaves earlier.

- Choose young leaves about 8 inches long, and cut them to within an inch of the ground.

- Eat whole young leaves in salad or cooked. The center stem on mature leaves can be tough, so cut it out.

Varieties to grow

Fordhook Giant, Bright Lights, Five-Colored Silver Beet, Rhubarb, Orange Fantasia, Golden, Flamigo

TOMATOES

(LYCOPERSICON ESCULENTUM)

How to grow

- Tomatoes are either determinate—a plant grows to a determined size, then flowers, sets, and ripens fruit all at about the same time— or indeterminate—a plant continues to grow vines and flowers throughout the season.

- Start seeds indoors 6–8 weeks before your last frost date. Plant 2 or 3 seeds ¼-inch deep in each 4-inch pot. Plants will emerge in 6–14 days. Thin to 1 plant per pot when the first set of true leaves appears.

- A warming tray (but not high air temperature) and strong lights help ensure success for indoor seedlings.

- Harden-off young plants by placing them outside during the day for a week. Then plant them in the garden or in containers outside.

- Plants in the garden should be 3–4 feet apart. Tomatoes are advantageous rooters, meaning they will make roots along their stems. Pinch off lower leaves and bury stem deeply, so that it can develop a healthy, extensive root system.

- Determinate tomato plants will need stakes to support fruit laden branches. Indeterminate tomato plants will need a substantial trellis.

- In cool climates, tomatoes benefit from season extenders—makeshift or permanent covers or protection to help raise the ambient temperature.

- Maintain soil moisture, especially in containers, to reduce the chance of blossom-end rot.

- Tomato lovers will say that you cannot plant too many tomatoes, but it helps to consider what you like best: Cherry tomatoes, slicers, or making fresh tomato sauce or paste to freeze.

- Encourage ripening by pruning leaves and branches that don't produce fruit. Remove all flowers that have not set fruit by mid-August. Reduce the amount of water you give plants so they focus on producing seed (inside the fruits) by being slightly stressed and abused.

How to harvest

- The blossom end (bottom) of ripe tomatoes should feel firm but slightly yielding and fruit breaks off the stem easily when snapped or twisted.

- Tomatoes ripen by the beginning of July in warm climates or August in cool-summer climates.

- Harvest indeterminates, such as cherry tomatoes, almost daily to keep up.

- At the end of the season, snip clusters of almost-ripe tomatoes, keeping them on the vine, and place them on the kitchen counter where they will continue to ripen

- If you have a big crop of green tomatoes at the end of the season, they can be used in relish, chutney, or muffins. Keep green tomatoes on the vine and check their ripening progress weekly—the bottom end will have a slight give when pressed.

Varieties to grow

The choice can be overwhelming, so do some taste-testing at your local farmers' market to find out which ones you like.

CHERRY: Sun Gold, Sweet Million, Black Cherry, Peacevine, Isis Candy

SLICING: Stupice, Oregon Spring, Black Prince, Odessa, Glacier, Siberian (all early producers) Celebrity, Momotaro, Brandywine, Garden Peach, Green Zebra, Mr. Stripey, Cherokee Purple

PASTE: Viva Italia, San Marzano, Amish Paste, Principe Borghese (for sun-dried tomatoes), Speckled Roman, Polish Linguisa

BASIL

(OCIMUM BASILICUM)

How to grow

- Basil is a warm-weather annual.

- Start seeds or plant starts outside, only after the weather is warm—end of May through June.

- Basil grows best when the soil temperature is 70 degrees and nighttime temperatures don't dip below 50.

- Thin seedlings to 12 inches apart.

- Pinch off any flowers before they open so that the plant will continue to grow more leaves.

- Plants in containers will benefit from a weak application of high-nitrogen liquid fertilizer every few weeks.

- Basil likes it warm and benefits from a cloche or hoop house if you live in a cooler climate.

How to harvest

- Pinch or snip off leaves next to the stem. Wash carefully, because basil leaves are easily bruised.

- Harvest continually during the summer so the plant will continue growing.

- Basil is killed at the first light frost. Before that happens, harvest the whole plant (or bring your plants indoors under lights).

- Harvested basil can be dried or put in the food processor with olive oil and frozen in ice-cube trays as an addition to winter cooking.

Varieties to grow

COMMON VARIETIES: Genovese, Globe, Purple, Dark Opal, Sweet Italian, Red Rubin, Italian Large Leaf

SPECIAL FLAVORS: Lemon, Greek, Cinnamon, Siam Queen (Thai)

CHIVES

(ALLIUM SCHOENOPRASUM)

How to grow

- Chives, a member of the onion family, grow from small bulbs. Chives are perennial: The plants die back in winter and appear again in spring.

- Grow chives from seeds or from small plants.

- Plant seeds in early May, ¼ inch deep, 1 inch apart. Plants will appear in 7–14 days.

- Thin seedlings to 6–8 inches apart.

- A small clump of chives will increase in size. They can be divided by digging down straight through the middle and moving part of the clump to another area, placing them in a container, or giving them to a friend.

How to harvest

- Use chives fresh by cutting off a handful (or however much you need) at the base of the plant with scissors.

- Chive blossoms—small round clusters of lavender flowers—can be used in salads and as garnish, either whole or pulled apart.

- Use the flowers to make herb vinegar, which turns a pretty pink color.

- Chop the stems and mix them with cream cheese.

Varieties to grow

Common, Garlic or Chinese (with white flowers), Flat-Leaf

CILANTRO

(CORIANDRUM SATIVUM)

How to grow

- Cilantro is an annual that bolts quickly in hot weather.

- Plant cilantro every 2 weeks beginning in late April into June, so that you will always have some leaves to use.

- Sprinkle the seed over the soil and cover the seeds with a light layer of soil. Keep watered, and the plants will appear in 10-16 days. Allow 3–4 inches between plants.

- If your winter is mild, sow cilantro in August for a fall harvest and a bit later to overwinter. Overwintered cilantro starts to grow early in the spring and is slower to bolt. You may get more than three harvests from overwintered cilantro.

How to harvest

- Start using when the plants are 3–4 inches high.

- Using scissors, give the tops a haircut; this will encourage more growth.

- Let a few cilantro plants flower and go to seed—and you'll have coriander. The seeds can be used fresh in stir-fry and salad dressing, dried for planting again, or used in pickling.

Varieties to grow

Confetti, Santo, Jantar, Vietnamese (Polygonum odoratum), Slow Bolt

How to grow

- A Mediterranean native perennial that thrives in hot sun, low humidity, and extremely well-drained soil.

- The easiest way to grow oregano is to buy a small plant.

- Grows well in containers, especially cold-hardy terra-cotta; hardy to zone 5.

- Dies back in winter, it can be cut back to the ground before spring.

- Can be grown as an annual; the entire plant can be harvested just before cold temperatures set in.

How to harvest

- Cut nonflowering stems early in the morning, tie up in bundles, and dry in a dark, airy room. When the leaves are dry, strip them off the stems and store the leaves in a bag or jar.

- Cut individual stems to use fresh; strip the leaves off by hand.

- Oregano flowers in small purple domes at the tops of stems. Use flowers as confetti in salads or in cut flower arrangements. It may reseed freely if the flowers are left to turn into seeds.

Varieties to grow

Common, Italian, Greek, Golden Marjoram is a sweeter, subtler cousin to oregano—grow Sweet or Golden Upright varieties.

(ORIGANUM VULGARE)

OREGANO

PARSLEY

(PETROSELINUM CRISPUM)

How to grow

- Parsley is a biennial—the first year it grows and the second year it wants to flower and set seed (and then die). Many gardeners grow it as an annual.

- Parsley seed is slow to germinate, so soak it in warm water the night before planting. Plant in the garden in May, setting the seeds ½ inch deep. The plants will emerge in 21–28 days.

- Thin to 6 inches apart once the plants are a few inches high.

- You may buy parsley plants at the nursery for your garden or containers.

- Parsley develops a long taproot, and grows best in loose soil or in a tall container.

How to harvest

- Cut leaves off as you need them.

- Dry parsley at the end of the season for winter use: Spread in a single layer on newspaper or cardboard out of direct sunlight. Store in an airtight container.

- The root of Hamburg parsley (a relative of the herb) is eaten: Grown from seed, it's ready to harvest in 3 or 4 months. Roast like a parsnip.

Varieties to grow

Italian (flat-leaf), Curly, Japanese, Hamburg (root)

ROSEMARY

(ROSMARINUS OFFICINALIS)

How to grow

■ A shrub native to the Mediterranean that thrives in hot sun, low humidity, and extremely well-drained soil.

■ The easiest way to grow rosemary is from a small plant.

■ An evergreen, hardy to zone 8, it is often seen as an aromatic, edible topiary (trimmed into a fancy shape) or as a hedge.

■ Purple or blue flowers appear on new growth.

■ Can be grown in containers as an annual and harvested before your first frost.

How to harvest

■ Harvest stems as needed for the kitchen or grill. Leaves can be stripped off the stem and chopped.

■ Stems can be bundled and dried in a dark, airy room. Remove dry leaves from stems and store in a bag or jar.

■ Rosemary flowers are great to graze on when you are out in the garden. They are mild with a slightly sweet, rosemary-flavored nectar.

Varieties to grow

Arp, Blue Spire, Tuscan Blue, Pink Flowered, French

SAGE

(SALVIA OFFICINALIS)

How to grow

- A shrub native to the Mediterranean that thrives in hot sun, low humidity, and extremely well-drained soil.
- The easiest way to grow sage is from a small plant.
- Deciduous, it drops most or all its leaves in winter. It's hardy to zone 4.
- Can be grown as an annual in containers.

How to harvest

- Harvest stems as needed for the kitchen or grill. Leaves can be stripped off the stem and chopped.
- Branches can be cut during the summer and dried in a dark, airy room. Strip stems of dry leaves and store leaves in a bag or jar.
- Leaves can be made into sage fritters or fried in butter.

Varieties to grow

Golden sage (Aurea), Tricolor, Purpurea, Italian Aromatic, Berggarten

THYME

(THYMUS VULGARIS)

How to grow

- A shrub native to the Mediterranean that thrives in hot sun, low humidity, and extremely well-drained soil.

- The easiest way to grow thyme is from a small plant.

- A evergreen small shrub with tiny leaves hardy to zone 5. Grows well in a container.

- Can be grown as an annual; the entire plant can be harvested in late fall.

How to harvest

- Harvest stems as needed for the kitchen or grill. Leaves can be stripped off the stem or entire stems can be put into soup or stew (and removed before serving).

- Branches can be cut during the summer and dried in a dark, airy room. Strip stems of dry leaves and store leaves in a bag or jar.

- Tiny clusters of pink thyme blossoms are great as a garnish or added to an herb bouquet.

Varieties to grow

English, German Winter, de Provence, Lemon, Silver, Orange Balsam, Caraway

Nonedible groundcovers: Creeping, Wooly

When you start your flowers from seed,
you are certain that they're grown organically.
If you purchase plants, make sure you buy from
a reliable source where chemicals have not been used.
Edible flowers are often used as a garnish on plates—
making a picture-perfect setting.

FLOWERS

How to grow

- Bachelor's buttons are annual flowers that come in white, purple, or shades of blue.

- Plant bachelor's buttons in spring in pots or directly in the garden.

- Left to set seed, these flowers will come up next year in your garden on their own. Bachelor's buttons can become invasive.

How to harvest

- Pick individual flowers and take only the petals, which have a mild flavor, for eating.

- Bachelor's buttons hold up well in cut arrangements and come in a variety of colors.

Varieties to grow

Black Gem, Choice Mix, Blue Boy, Black Ball

(CENTAUREA CYANUS)

BACHELOR'S BUTTONS
AKA CORNFLOWERS

How to grow

■ Borage is an annual plant with fuzzy stems and leaves which produces clusters of small blue flowers with black centers—sometimes a plant will have blue and pink flowers. The whole plant smells and tastes lightly of cucumber.

■ Grow borage from seed planted in the garden in late April.

■ Bees and other pollinators love borage flowers, so look before you pick!

■ Borage seeds are important food for small songbirds, so let your plants set seed to support urban wildlife.

How to harvest

■ Pick the flowers for salads or to decorate cheese plates.

■ Freeze the flowers in ice cubes for summer drinks.

■ Borage will reseed in your garden to reappear next spring but will not become invasive.

Varieties to grow

Blue (most common), White (rare)

BORAGE

(BORAGO OFFICINALIS)

Calendula

(CALENDULA OFFICINALIS)

How to grow

- Calendula is an annual plant with daisylike flowers that grows best in cool-weather climates. The flowers have a slightly bitter but not disagreeable taste.

- Grow calendula from seed by planting in early spring ¼-inch deep.

- Calendula can also be grown from plants bought at a nursery.

- Calendula will flower in early summer (spring in some mild-weather areas) and again in the fall.

- Calendula may reseed itself, and you will need to plant it only once.

- Calendula is the gardener's friend; it grows easily, reseeds itself, but doesn't become invasive. It is edible, and infused oil will help scrapes, bites, or sunburn heal more quickly.

- In mild climates calendula will flower every month of the year.

How to harvest

- Pick the daisylike flowers one by one and pull off the petals to sprinkle on salads or sandwiches.

- Flowers are good in cut arrangements.

- Dry the petals in a single layer on paper. Dried calendula petals can be ground and used in cooking as a substitute for saffron.

- Grind dried calendula petals with dried comfrey leaves and lavender blossoms and combine this mixture with baking soda for a homemade anti-itch powder.

Varieties to grow

Calendula comes in shades of orange, apricot, yellow, and cream. Flashback petals are dark orange underneath—very striking!

Pacific Beauty Mix, Neon, Déjà vu, Triangle Flashback, Flashback Mix, Sunshine Flashback

DIANTHUS

(DIANTHUS)

How to grow

■ Dianthus (sometimes called pinks or rock carnations) are perennial plants that flower in early summer in white and shades of pink and red. The flowers are fragrant and have a spicy-sweet flavor.

■ Grow dianthus from plants purchased from an organic grower.

■ Plant in a hot, sunny spot; dianthus prefer alkaline soil, so plant near concrete stepping-stones or a sidewalk.

■ Trim off any unused flowers at the end of the summer to neaten up the plant.

■ Dianthus are hardy in zones 3–8. Grow as an annual in colder regions.

How to harvest

■ Pick the flowers and use the petals as a garnish or in salads.

■ Flowers make a cute bouquet with a nice fragrance.

Varieties to grow

Chianti, Fire Witch, Pink Lace

How to grow

- Nasturtiums are a vining or clumping annual plant with flared flowers (like a trumpet) that come in shades of orange, red, yellow, and peach. Leaves are round and scalloped; some varieties have variegated leaves. The leaves and flowers have a peppery flavor.

- Plant nasturtium seeds in April in pots or in the ground about ½ inch deep.

- Although nasturtiums do not need extra fertilizer, they do best in average soil and with regular water.

- Nasturtiums will self-sow but will not become invasive.

How to harvest

- All parts of a nasturtium are edible and eaten fresh most of the time. Pick the leaves and flowers for salad or sandwiches.

- Nasturtiums are a magnet plant for aphids, so pick flowers regularly and inspect them well before eating.

- Sip the sweet and spicy nectar by nibbling off the end of the flower spur and sipping the tasty juice.

- Use the flowers to make herb vinegar, which turns the same bright shade as the flower.

- Harvest the seeds and pickle them to use as a caper substitute.

Varieties to grow

Empress of India, Dwarf Jewel Mix, Cherries Jubilee, Alaska Mix, Spitfire, Whirly Bird, Creamsicle, Moonlight, Vanilla Berry

NASTURTIUMS

(TROPAEOLUM MAJUS)

How to grow

- Sunflowers are an annual known for their tall stalks and huge flowers, although they come in shorter, multistemmed varieties too. Flower colors include cream, bronze, maroon, orange, and yellow; centers can be dark or light.

- Plant from seed in April or May, when soil temperature is above 75 degrees. Plants will emerge in 7–14 days.

- Thin plants to 12 to 18 inches apart.

- Plant in paper pots and grow inside or in a cold frame to get an earlier start.

- Stake the giant, heavy-headed varieties or grow them up against a fence where their heads can be supported as they grow.

- Sunflowers have an allelopathic quality, meaning that they inhibit the growth of other plants. Give sunflowers their own bed and grow a thick stand that will put on a fantastic show and serve as a cool hiding place!

How to harvest

- When the leafy backside of the flowers begins to shrivel, cut the flowers off with enough stem so that you can hang them upside down in the garage or storeroom to dry.

- Fresh seeds can be peeled and the sweet, soft kernels eaten right from the maturing flower.

- Roast sunflower seeds for yourself or offer them to the birds.

- The unopened buds of sunflowers can be cooked like an artichoke.

- Save dried seeds to plant again.

Varieties to grow

There are so many sunflowers available that listing varieties would be a disservice. Pick a couple of varieties that strike your fancy and enjoy the show!

(HELIANTHUS ANNUUS)

SUNFLOWERS

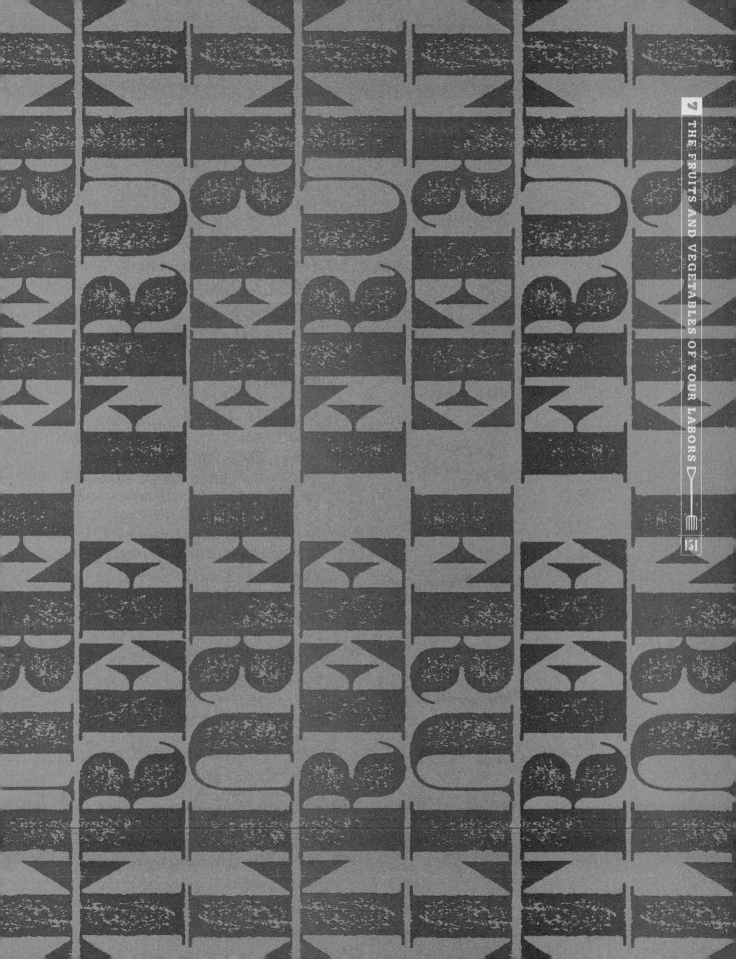

APPLE

(MALUS)

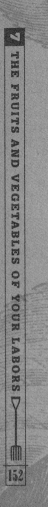

How to grow

- Apples grow best in zones 4–9.

- Apple trees can grow 20–40 feet high and wide; choose a columnar variety or one that grows on a dwarf or mini-dwarf rootstock for small spaces or large containers.

- Espaliered apples are grown in a two-dimensional form in the shape of a fan or fence and take up less room than a full-size tree.

- Numerous apple varieties can be planted as a cordon fence.

- For best fruit set, grow 2 kinds of apples that bloom at the same time or a crab apple and an eating apple. You can also plant a 2- or 3-in-1 tree with more than one variety on the same trunk.

- Plant in fall or winter as potted plants or in early spring when apple trees are available as bare rootstock.

- Look for a list of disease-resistant varieties for your region from your local cooperative extension service; there you will also find information on pruning your apple tree.

How to harvest

- Some apple varieties begin ripening as early as mid-August; some as late as the end of October.

- An apple is ripe when it breaks off the stem when you twist and pull.

- Cut the apple in half and look at the seeds; an apple is ripe when the seeds are black.

- Early apples tend to have a short shelf life; later-ripening apples are better keepers.

Tree rootstock for city farms

Columnar apples: M7, M26

Mini-dwarf: M9, M27, P22, EMLA 27

Semidwarf: M7, M26, M106, EMLA 26, EMLA 111

How to grow

■ Apricots grow in zones 4–9.

■ Choose an apricot on a dwarfing rootstock, such as Mar2624 or Lovell, for small-scale city farms.

■ Many apricots are self-fertile, but most set fruit better when another variety of apricot that blooms at the same time is grown nearby.

■ Thin fruit clusters to increase air circulation and avoid brown rot (dusty thick mold that covers fruit).

How to harvest

■ Apricots ripen from June through August, depending on the variety and on the climate.

■ Fruit should be easy to pluck from the tree when ripe.

■ Eat apricots fresh or preserve them by canning, freezing, or drying.

Varieties to grow

Puget Gold, Harglow, Perfection

APRICOTS

(PRUNUS ARMENIACA)

CORDON

ESPALIER

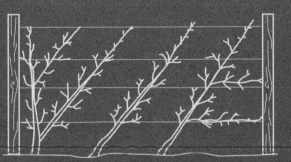

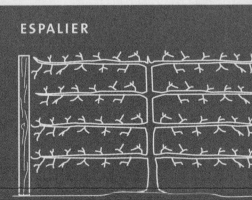

CHERRY

(PRUNUS SPP.)

How to grow

- Sweet cherries grow in zones 5–9; sour (also called pie or tart) cherries grow in zones 6–10.
- Choose a cherry variety (sweet or sour) grown on Gisela-5 (G-5) rootstock. The tree will grow about half its normal height, making it easy to include in a city farm and also making it easy to net against birds
- Sour cherries are all self-fertile and don't need a second tree to pollinate them. Some sweet cherries are self-fertile, while others need a different variety to pollinate. Check carefully before purchasing to make sure you get what you want.

How to harvest

- If netted for birds, you won't have to worry about birds getting to your cherries first.
- Fruit-picking season varies, depending on the variety, and can span the whole month of June. Sour cherries ripen later than sweet ones.
- Pick cherries by twisting the slender stem and pulling, being careful not to tug the stem from the fruit, which can damage the fruit.
- Eat sweet cherries fresh or preserve sweet or sour cherries by canning them in syrup, freezing, or drying.

Varieties to grow

Sweet, self-fertile: Lapins, Sweetheart, Black Gold, Stella, White Gold

Sweet, not self-fertile: Bing, Early Burlat, Hartland

Sour cherries are self-fertile: English Morello, Surefire, Montmorency, Danube

Peaches and Nectarines

(PRUNUS PERSICA)

How to grow

- Peaches and nectarines grow in zones 5–9.

- Choose a tree on a dwarf rootstock, such as Lovell or Pumi Select, for small-scale farms.

- Peaches are available in combinations—several varieties growing on one tree.

- Peaches and nectarines need warmth and do best in climates with hot summers; there are a few varieties that have been bred for cool-summer climates.

- Thin fruit clusters to increase air circulation and avoid brown rot (dusty thick mold that covers fruit).

How to harvest

- Peaches and nectarines are ready to harvest beginning in July in hot-weather regions and August in cooler climates.

- Ripe fruit should come easily from the tree when plucked.

- Peaches and nectarines are highly perishable. Eat fresh, can, or freeze (fruit can be cut in half or sliced to freeze).

Varieties to grow for cool-summer regions

Peach: Contender, Early Loring, Harken, White Lady

Nectarine: Hardired, Arctic Jay White,

How to grow

- Asian pears, which are crisp as an apple but taste like a pear, grow best in zones 5–9.

- Asian pear tree varieties are available on semidwarf rootstock; trees will reach about 15 feet high and wide.

- For best fruit set, grow two kinds of pears that bloom at the same time. Or buy a 2- or 3-in-1 tree that has more than one variety grafted on the same trunk.

- Plant in fall or winter as potted plants or in early spring when Asian pear trees are available as bare rootstock.

- Look for a list of disease-resistant varieties for your region from your local extension office; there you will also find information on pruning your Asian pear tree.

- Asian pears on semidwarf rootstock can be espaliered.

How to harvest

- Asian pears are harvested when ripe. Check the ripening time for your variety; then test a pear every few days to see if it's time.

- Asian pears can be stored in the refrigerator for several weeks.

Varieties to grow

European semidwarf: OHxF97, OHxF333
Asian semidwarf: OHxF97

PEA
Asian

How to grow

■ Pears grow best in zones 4–9.

■ Pear tree varieties are available on semidwarf rootstock; trees will reach about 15 feet high and wide. Look for Old Home X Farmingdale (OhxF) rootstock.

■ For best fruit set, grow two kinds of pears that bloom at the same time. Or buy a 2- or 3-in-1 tree that has more than one variety grafted together.

■ Plant in fall or winter as potted plants or in early spring when pear trees are available as bare rootstock.

■ Look for a list of disease-resistant varieties for your region from your local cooperative extension service; there you will also find information on pruning your pear tree.

■ Semidwarf pears can be trained as a cordon or espalier.

How to harvest

■ Pears can be picked from late summer into fall, depending on the variety.

■ Pears should be picked just before they are ripe, and allowed to ripen on the counter or in the refrigerator. Pears should be firm and green when picked; seeds will be black.

European

PEARS

(PYRUS SPP.)

PLUM

(PRUNUS)

How to grow

- Plums grow in zones 4–9.

- Choose a plum tree on semidwarf rootstock (Marianna 2624 or Krymsk 1) to fit into your city farm.

- Some plum trees are self-fertile, but others need another tree to help set fruit.

- Grow a plum tree grafted with several varieties, and you will be sure to have good cross-pollination and a fine selection.

- If you live in a climate with a late frost date, choose plum varieties that bloom late.

- Thin fruit clusters to increase air circulation and avoid brown rot (dusty thick mold that covers the fruit).

How to harvest

- Depending on the variety, plums begin ripening in July.

- Pluck the plum from the tree to harvest; when ripe, it should come away from the branch easily.

- Plums can be eaten fresh or preserved by canning, freezing, or drying.

Varieties to grow

Self-fertile: Victoria, Italian, Hollywood, Methley

BLUEBERRY

(VACCINIUM)

How to grow

- Blueberries are deciduous shrubs that grow in zones 3–8.

- Include blueberries in the landscape for their beautiful fall color as well as their delicious fruit.

- Blueberries prefer acidic soil; grow near other acid lovers, like rhododendron or conifers. When grown as a hedge or in a group, it is easier to maintain the proper soil pH.

- Choose small-growing varieties for city farms with limited space; they can be grown in containers—water frequently during fruit development.

- You may be able to increase your harvest by planting blueberries that ripen at different times—early, mid-, and late season.

How to harvest

- Blueberries are red when they are green—in other words, don't pick them until they are dark blue!

- Blueberries are ready to harvest beginning in early July through August, depending on the variety.

- Pick blueberries over many days, so that you can get the ripest fruit each day. Berries should come away from the plant easily.

- Eat fresh or freeze them. Make blueberry jam or bake them in a pie.

Varieties to grow

Small: Northsky, Chippewa, Top Hat

Early: Patriot, Spartan

Midseason: Bluecrop, Bluegold

Late season: Darrow, Jersey

Evergreen: Sunshine Blue

CURRANTS

(RIBES)

How to grow

- Currants are medium-sized deciduous shrubs that grow in zones 3–8.
- Grow red or black currants.
- Currants are self-pollinating.
- Some states restrict the sale of currants from mail-order sources because the shrubs are an alternate host to rust disease that affects North American native white pines.

How to harvest

- Currants ripen in late June or early July.
- It's tedious to pick each fruit, so "comb" down a branch, holding a bowl underneath. You'll get some leaves, too, but you can pick those out in the kitchen.
- Currants are most often juiced, used in preserves, or dried like raisins.

Varieties to grow

Red: Jhonkheer Van Tets, Red Lake, Cherry

Black: Titania, Ben Sarek, Ben Lomond, Boskoop Giant

White: Blanca

ELDER OR ELDERBERRY

(SAMBUCUS SPP.)

How to grow

■ Elders or elderberries are large deciduous shrubs that grow in zones 4–9.

■ Elderberries are toxic when eaten fresh.

■ Elderberries grow into large shrubs with flat, umbrella-shaped clusters of white or pinkish flowers that bloom in May. Clusters of blue or black berries ripen in summer.

■ Elderberries make good hedge plants, grown with other shrubs for ornament and food.

■ Cut elders back in winter to keep their growth more compact.

How to harvest

■ The elder is an ancient medicinal plant. You may use the flowers (fresh or dried) in tea and the berry juice in syrup, jam, or wine.

■ Elder flower cordial: Cut about 15 elder flower stems and steep them in sugared lemon water for 48 hours—use the rinds and juice of two lemons, 24 ounces of boiling water, and 1 pound of sugar. Strain and use the liquid as a cordial, to drink over ice with club soda or gin.

■ Cut stems of berry clusters in summer when fruit hangs heavy. Elder stems are not edible and the skin and seeds of berries are mildly toxic. Cook berries (removed from the stems) and strain. Use as juice, syrup, or make into wine.

Varieties to grow

Sambucus nigra: Allesso, Black Lace, Black Beauty, Variegated, York, Nova

Sambucus Caerulea: Blue Elder

How to grow

■ Figs are large, deciduous shrubs or small trees that grow in zones 7/8–9/10.

■ Figs make good ornamental edible plants, and a few small varieties can grow in containers.

■ In cooler climates, grow figs up against a south-facing wall for more heat.

■ Most figs set two crops, one in the fall that ripens in spring (those figs do not ripen in zones 7 and 8), and a big crop in spring that ripens in fall.

■ Figs do not need to be cross-pollinated.

How to harvest

■ Harvest figs in the late summer and early fall by lifting a fruit, twisting, and tugging; it should come off easily.

■ Figs can be eaten fresh or preserved by freezing, drying, or canning in syrup or brandy.

Varieties to grow

Brown Turkey, Desert King, Mission

FIG

(FICUS CARICA)

How to grow

■ Grapes grow as a deciduous vine in zones 5–9, and produce tendrils that twine around supports.

■ Grapes produce best when trained onto a fence where branches can grow laterally.

■ Grapes need to be pruned hard in winter. Check with your local extension service for a diagram on pruning methods.

■ Grape varieties are usually divided up into table grapes (for fresh eating) or wine grapes.

■ The type of grape you grow depends on the number of "heat units" your region gets.

How to harvest

■ Harvest grapes when they are fully ripe; test a single grape every few days for flavor.

■ Grapes are harvested in the fall by cutting off bunches with a knife or pruning shears.

■ Eat fresh, make into juice or jam, or dry the fruit for homegrown raisins.

Varieties to grow

Check with your local extension service to find out which varieties grow best in your area.

GRAPES
(VITIS)

STRAW-BERRIES

(FRAGARIA)

How to grow

- Depending on the variety, strawberries produce one big crop, usually in June (once-bearing), produce small amounts of fruit throughout the summer (day neutral) or produce two crops, one early and one late in the summer (ever-bearing).

- Alpine strawberries produce small, slender, sweet fruit on compact plants that can be incorporated into the ornamental garden as a groundcover; they will grow in part shade.

- For vigorous, disease-free plants, buy them from the nursery.

- Plant 15 inches apart in slightly raised rows, which are 36 inches apart, and when runners appear, train them to grow in the spaces between rows (cut off some so that the area doesn't get too crowded).

- Plants should be buried so that the roots are belowground, but the growing point (the center where leaves emerge) is above soil level.

- Strawberries can be grown in "strawberry pots"—terra-cotta containers with pockets on the side—but care must be taken that the plants do not dry out.

How to harvest

- Strawberries are ripe when they are fragrant and red (the shade depends on the variety).

- Pick strawberries fully ripe by holding the berry and breaking the stem off.

- Eat strawberries fresh or preserve them by freezing or making jam.

Varieties to grow

Once-bearing: Earliglow, Allstar, Shuksan, Firecracker

Day neutral: Tristar, Seascape, Everbearing

Alpine (Fragaria vesca): Rugen, Mignonette, Yellow, White

How to grow

- Raspberries and blackberries are cane growers that will increase their stand by growing new canes from the ground. Canes grow from 4 to 6 feet high, and can be trained onto a fence.

- Raspberries and blackberries come in varieties that produce one summer crop or those that produce crops in summer and in September (ever-bearing).

- For once-bearing varieties, cut the canes that produced fruit to the ground in September. New canes will produce fruit the next year.

- Ever-bearing types can be cut to the ground in late winter/early spring. New canes will produce one large, late (August/September) crop.

- Choose a thornless blackberry variety for easy picking family fun!

How to harvest

- Raspberries and blackberries ripen in July (or August/September for ever-bearing varieties).

- Pick every 2 or 3 days to make sure no berries are lost.

- Ripe raspberries will come away easily, and leave a white stalk behind (and a hole in the berry); blackberries should taste sweet and come away easily (white core will stay with fruit).

Varieties to grow

Blackberry: Marionberry, Tayberry, Boysenberry, Thornless Loganberry, Triple Crown Thornless, Black Pearl Thornless, Arapaho Thornless

Raspberry: Boyne, Latham, Meeker, Tulameen, Cascade Delight, Saanich

Ever-bearing Raspberry: Summit, Autumn Bliss, Caroline, Rosanna, Autumn Britten

Raspberries and Blackberries
(RUBUS)

The Row to Hoe

There are so many different vegetables and fruit to grow that it can be difficult to narrow it down. My advice is to

BEGIN WITH A FEW THINGS YOU LOVE TO EAT.

add a berry vine or a fruit tree for homemade jam; and try planting a few rare or endangered vegetable varieties. Share what you grow and what you learn about city farming like seeds in the wind.

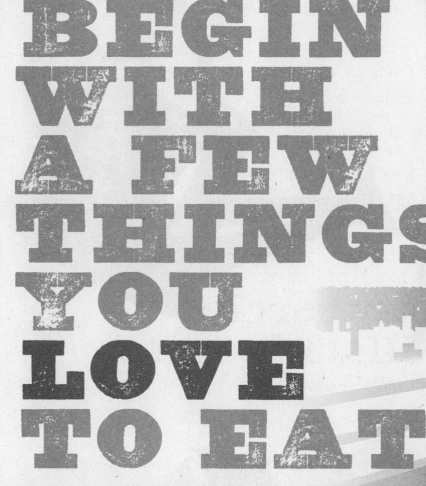

KEEPING IT GROWING

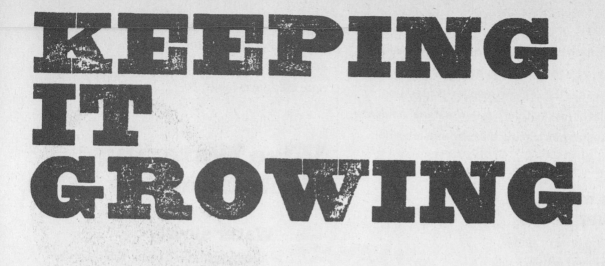

ATER IS ONE OF THE LARGEST expenses for the city farmer. Watering wisely conserves water, saves money, and promotes both healthy plant growth and bigger yields. In this chapter you'll learn how much to water, how to choose the right watering tools, and which techniques make every drop count!

You can put your garden in the right place and build healthy soil, but if you forget to water or do it inconsistently, you'll end up with disappointing yields. Vegetables are 80–95 percent water and require consistent amounts of water to grow. Your job is to keep the soil near the roots evenly moist so the plants can take what they need. Without consistent water, you will have later harvests, lower yields, and small, dense vegetables with a bitter taste. Getting water to your garden is one of the biggest jobs for the city farmer.

Ways to Conserve Water

Build healthy soil

By adding organic matter and compost to dirt, you are increasing your soil's sponge action. Organic matter soaks up water and then slowly releases moisture to plants and soil organisms. Healthy soils also filter storm runoff. Compost slows drainage in sandy soils and helps loosen clay soils so they absorb water better.

Mulch

Any irrigation plan will be improved by the use of mulch all year round. As you have already learned, mulch shades the soil and reduces evaporation. Due to its porous nature, organic material holds water, thereby increasing the moisture retention of your soil.

Drip irrigation or soaker hoses

Using drip lines or soaker hoses applies water slowly, allowing it to seep deeply into the soil, where it evaporates slowly and is accessible for plants to use. Drip systems keep water on the soil's surface rather than on plant leaves; this reduces water lost to evaporation and curbs disease transmission.

Grouping plants

Clustering plants according to water needs makes sense and makes your work easier. Planting new plants that require frequent watering in one area focuses your watering efforts. Group older, more established plants that don't require much water together. Put drought-tolerant plants on the outer edges of your property, since they require little care once they are established.

Wise Watering Techniques

Water slowly

Ideally, you would water each plant as though you were pouring a cup of tea—carefully pouring just the right amount of water so that the soil absorbs it as you pour. Seedbeds and seedlings would be misted with a gentle shower from an English watering can. This isn't practical, but it is an excellent way to think about applying water slowly. Much of the water used in lawns and gardens runs off before it makes it to the roots. Allow water to soak deeply into the soil; dig down to check how deep water permeates. If puddles occur, turn off your hose and let soil absorb all the water before continuing.

KIDS + WATER

Water less frequently and deeper

Encourage healthy, strong root growth by letting the top 1 or 2 inches of soil dry out between waterings. Allowing the soil to dry out a little between waterings will encourage roots to penetrate deep into the soil, seeking out moisture.

This will promote healthier and more extensive root systems. Healthy plants are more resistant to disease and more resilient in the face of extreme temperatures, withstanding drier, hotter conditions. The roots of plants grow toward moisture. If you keep only the top inches moist, the root system will be shallow.

Water consistently

Water vegetables the same amount or length of time at steady intervals.

WATERING TIPS

- *Keep the soil around the roots of your plants evenly moist. To find out if your plants need water, dig down to see if the soil is damp and dark.*
- *Seedlings require more frequent watering than mature plants.*
- *Morning is the best time of day to water.*
- *Water the soil, not the plant.*
- *Water deeply and less frequently. Keep water pressure low so that water soaks into the soil instead of running off the surface.*
- *Conserve water by mulching, adding compost, and grouping vegetables with similar water needs.*

For example, you might water your garden for 10 minutes every two days. If you miss a day or two of watering (or get off your schedule), don't try to make up for it by overwatering. Inconsistent watering is the main cause of blossom-end rot and split or misshapen fruit.

Water in the morning

Water droplets on leaves near the soil encourage the growth of fungal diseases, such as late blight and powdery mildew. Watering before 10 a.m. gives the plants time to dry off their leaves, reducing the chance of disease transmission. Watering in the evening is okay, but cools the soil (bad for heat crops), attracts slugs, and may promote disease. Regardless of the time of day, if plants are badly in need of water, give it to them!

INSTANT FUN!

When we teach kids how to water plants in Seattle Tilth's Children's Garden, we liken it to taking a drink of water from a cup. When you drink water, you don't just open your mouth and throw water at it, hoping that some gets inside. If you drink too fast, you might choke or get water all over your shirt. Instead, you carefully sip and swallow.

No need to buy cutesy watering cans for your children. Use recycled yogurt cups. They are easy to fill from a bucket, and will fit your child's hands. Have kids water plants as though they are drinking. Give the plants a sip and then let the soil absorb the water (swallow). Sip and swallow.

Keep your garden weeded

Weeds are greedy and compete with crops for water. Use mulch or drip irrigation to reduce weed-seed germination. Pulling weeds is easier when the soil is moist. Chop annual weeds into the soil before they have set seed to help maintain soil fertility.

Set timers

Use a simple twist-dial or battery-operated timer to control how long sprinklers or irrigation systems run. That way if you get sidetracked harvesting or weeding, you won't risk overwatering.

Repair leaks

Keep faucets and hoses in good working order so that water is always available for plants. Large volumes of water are wasted as a result of leaky faucets and hose couplings. Leaky couplings can easily be replaced with high-quality parts that will last for years.

How Much to Water

The Right Way to Water Your Plants

- *Seedbeds must stay moist until germination. Then they are watered less frequently as seedlings grow.*

- *New transplants or seedlings will need daily water for the first week after transplanting. They will need between 1 and 2 gallons per plant each week, as they mature.*

- *Fruit trees will need more water as flowers and fruit develop and less water as fruit ripens.*

- *Drought-tolerant plants and perennial herbs need consistent water until they are established (2 or 3 years). These plants may need water only 2 or 3 times each summer after they are established.*

The root depth of plants affects how much you will need to water them. Shallow-rooted plants may need frequent watering, since the top few inches of the soil dries out quickly in hot weather or in sandy soils. Deeper-rooted plants are more drought-tolerant because they can tap into water and nutrient reserves deep in the soil.

Root Depth of Vegetables

SHALLOW (12–18")	MEDIUM (18–24")	DEEP (more than 24")
broccoli	beans	asparagus
Brussels sprouts	beets	okra
cabbage	eggplant	parsnips
carrot	mustard greens	pumpkin
celery	kale	rhubarb
Chinese cabbage	lettuce	sweet potatoes
collard greens	peas	tomatoes
corn	peppers	watermelon
cucumbers	potatoes	winter squash
leeks	rutabagas	
onions	summer squash	
radishes	turnip	
spinach		

Water Needs of Common Vegetables

FREQUENT	MODERATE	LOW
broccoli	beans	artichoke
cauliflower	beets	asparagus
celery	Brussels sprouts	okra
Chinese cabbage	cabbage	parsnips
cucumbers	carrot	sweet potatoes
lettuce	collard greens	
mustard greens	corn	
onions	eggplant	
peas	leeks	
radishes	peppers	
spinach	potatoes	
summer squash	pumpkin	
	rhubarb	
	rutabagas	
	tomatoes	
	turnips	
	watermelon	
	winter squash	

Critical Watering Times

CONTINUOUS	ROOT & BULB EXPANSION	LEAF & HEAD DEVELOPMENT	FLOWERING & FRUIT DEVELOPMENT
celery	beets	broccoli	beans
Chinese cabbage	carrots	Brussels sprouts	cantaloupe
collard greens	parsnips	cabbage	cucumbers
kale	rutabagas	cauliflower	eggplant
leeks	turnips	lettuce	okra
mustard greens	artichoke		peas
radishes	asparagus		peppers
Swiss chard	okra		potatoes
	parsnips		pumpkin
	sweet potatoes		summer squash
			tomatoes
			watermelon
			winter squash

Vegetables Need Water

All vegetables need plenty of water, but some need more than others. Most vegetables have high or moderate water needs.

Important times to water

Once established, however, there are critical times for vegetables to get moisture. Some need water throughout their growth (that is, continuously); others need water as their roots or bulbs are forming and enlarging. Vegetables such as lettuce and broccoli need water as leaves and heads are developing and fruit-bearing vegetables need water during flower and fruit development. The chart "Critical Watering Times for Vegetables" gives you an idea of the most important times for getting water to different common vegetables.

How to Water

Clearly watering is plant-specific. Here are some guidelines for watering your different growing areas and plants.

■ **Seedbeds:** *Keep the top 2 inches of soil evenly moist to promote even germination. Water daily (or twice a day if your soil dries out quickly). Keep the soil moist, but do less watering as plants grow. Give them a drink every 2 or 3 days. Monitor soil moisture with your finger.*

■ **Seedlings or transplants:** *Water daily or every other day for the first 2 weeks, then every 2 to 3 days. To conserve water and give transplants the best start, don't plant in hot, dry weather. Transplant in the cool of the morning and water the soil well both before and after you plant. To reduce the need for lots of added water, transplant trees and perennials in early spring or in the fall when the weather is cool, water evaporates more slowly, and rain can help alleviate transplant shock.*

■ **Containers:** *Check your plants daily in the summer. Add enough water so that it drains out the hole in the bottom. Increase water retention by adding compost or coir to your potting soil mix.*

■ **Large herbs, shrubs, or trees:** *These need water, especially in dry hot weather, until they are established. Perennial plants need less water because of their more extensive root system. Water new trees 3 times a week for the first month after transplanting. Water once a week for the first year.*

Watering depends on your soil

Moisture availability to plants depends on the type of soil they're growing in and is an important consideration in planning your irrigation system. Sandy soils drain well and dry out quickly. Use less water more frequently so that your soil stays consistently moist down to the root zone. Clay soils need less water and more time for water to be absorbed—drip irrigation or soaker hoses are excellent choices for clay soils. Adequate moisture levels range from soil that feels nicely moist to soil that's slightly soggy.

Are Your Plants Getting Enough Water?

Feel the soil

One to two hours after watering, use a small trowel or hori hori to dig down and see how deeply water has penetrated. If soil doesn't feel moist at the root zone, water some more. If the soil is soggy, you have watered too much. Soil should look dark and feel moist, like devil's food cake.

Look at your plants

Pay attention. Watch for signs from your plants that they are not getting enough or are getting too much water. Some plants will naturally wilt or go limp during the heat of the day and will perk up when it cools off in the evening. Plants that are getting too much water will also look droopy and their leaves will turn yellow and wither. Before you head for the hose, check the soil to make sure it is moist but not soggy. If the soil is cool and moist, wait until evening or morning and water as usual. If it is soggy, stop watering for a few days and then reduce the

Pinpointing the Source of Water Problems

INADEQUATE WATER

SYMPTOMS	VEGETABLES
bitter or sharp flavor	lettuce, broccoli, radishes, cucumber
tough leaves	kale, collard greens, spinach
small and poorly filled pods	beans, peas, peppers
pithy or woody roots	radishes, beets, carrots, parsnips, turnips
pithy stems	rhubarb, celery
shriveled stems or fruit	asparagus, peppers, beans
misshapen or tiny roots	beets, carrots, radishes
misshapen fruit	tomatoes, peppers, squash

INCONSISTENT WATER

SYMPTOMS	VEGETABLES
blossom end of fruit turns yellow, then brown and starts to rot	tomatoes, peppers, squash, cucumber, eggplant
cracked fruit	tomatoes, cucumber, beets, carrots

TOO MUCH WATER

VEGETABLES	SYMPTOMS
all vegetables	leaves turn yellow, then wilt and wither
	no new leaf growth after yellowed leaves wither
	rotten roots
	stems are soft and mushy
	mold, algae, or moss grows on soil surface and on bottom leaves

amount of time you water. If your soil is bone dry to the root zone, water slowly, as if you were pouring a cup of tea, regardless of the time of day.

Ways to water

Watering tools get some of the heaviest use of all the equipment in your shed. You will use hoses and watering cans several times each week, all season long. Fancy, shiny watering tools are such a temptation at the garden store. That gleaming red water nozzle with 10 different spray patterns and squeeze trigger is so cool! Don't be lured by these tools. Water nozzles and sprinklers get dirt and moisture in cracks and crevices; after a season or two, parts seize up and stop working and can't be repaired. Stick with simple utilitarian designs that have few moving parts; these last a long time and can be repaired.

water nozzles

garden hose

Easy Water System Setup

A simple garden hose with a fan-shaped or round spray head, a shut-off valve, and a simple sprinkler is all many gardeners need to water their garden. For a great basic watering system, you will need a watering can, garden hose, ball valve, water nozzle, and sprinkler. Pick those up and you're all set! Remember to keep your water pressure low to allow water to soak into the soil slowly, rather than run off into paths or down the sidewalk.

watering can

Watering cans

These are great for seedbeds and seedlings when you want a fine gentle spray and a sprinkler would be too much. Many plastic watering cans are cheaply made and last only a couple of years before the end breaks and the water head becomes worthless. High-quality plastic watering cans have replaceable water heads and are well worth the investment. Galvanized metal English watering cans look great and are expensive, but will last a lifetime. High-quality plastic and galvanized watering cans offer water heads (or roses) with fine or medium spray. Rather than screwing on, roses fit snuggly over the can spout.

Garden hose with water breaker

Find a high-quality, reinforced, lead-free garden or boat/RV hose. Put a brass ball valve on the end of your hose so that you can adjust the water where you are, rather than running back and forth to the faucet. Ball valves with one large handle are easier to use than those with a button-style knob. Round or fan-shaped water spray heads come in different rose sizes, from mist to medium spray.

Alternative watering systems

There are many ingenious ways to reuse common items for watering your garden. Handmade drip irrigation gizmos can be created out of old buckets or gallon water jugs. Punch a hole in the side of a 5-gallon bucket or near the bottom of a water jug so that water dribbles slowly into the soil. This is a great way to fertilize and irrigate (fertigate) crops using liquid fertilizer. Use a 5-gallon bucket irrigator for outlying crops of pumpkins or artichokes.

ball valve

sprinkler

Sprinklers

There are many different kinds of sprinklers, look for metal (not plastic) sprinklers that have a set pattern and few moving parts:

- *Spike sprinklers—simple design, no moving parts, easy to move to water multiple beds, come in different spray angles.*
- *Other sprinklers—small sprinklers come in different spray patterns to match bed shape.*
- *Oscillating sprinklers—these classic sprinklers are great for watering large vegetable gardens or the lawn.*

Sprinklers water a wide swath, encouraging weed seeds to germinate and increasing your weeding. Use a timer on your sprinklers to avoid over watering. Adjust sprinklers so you don't overspray or water the driveway.

Plumbing and the city farmer

As soon as you attach a hose to your faucet, you become a plumber. Don't be scared. Plumbing doesn't need to be difficult or intimidating. The worst thing that can happen is that you'll spring a leak. Your job is to keep your hose or irrigation system leak-free. Plan on three trips to the hardware store for every plumbing project you undertake in the garden. If you complete your task with fewer visits, you are way ahead of the game!

The female coupling of all garden hoses start to leak over time. If your hose is leaking at the faucet, replace the rubber washer in the female coupling and see if that works. If the coupling is leaking from the swivel, you will need to either fix the hose or buy a new one.

Replacement Parts

Most of these parts will be grouped with garden supplies
at your hardware store; hose clamps may be located
in the general plumbing section.

1

2

3

4

1 Brass male and female couplings: Plastic couplings with plastic hose clamps are impossible to install and will leak. Save your money and choose brass couplings and simple metal hose clamps.

2 Hose clamps: Use $9/16$-inch to $11/16$-inch sized clamps for most garden hoses.

3 Rubber washers: Replace rubber washers every season.

4 Socket or combination wrench to tighten clamps: Slotted screwdrivers can slip as you are tightening the hose clamp and jab into your hand (I have a scar to prove it). A combo or socket wrench is a better choice.

How to replace a leaking hose coupling

2

4

3

1 Cut off the damaged section **2** Slip on hose clamp **3** Insert fitting **4** Tighten clamp

Fixing your hose

The fix is easy—all you need is a replacement brass female coupling and a hose clamp. Make sure the stem of your coupling is the same size as the inside of your hose. I take a little section of hose with me to the hardware store to make sure I get the right size on my first visit!

Cut off the old coupling with a sharp box cutter. Slide the hose clamp onto the hose. Push the stem of the new coupling into the hose (use a little dishwashing soap if you need a bit of lubrication). The stem should fit snuggly. Slide the hose clamp up so that it is near the base of the female coupling. Tighten the hose clamp (use a combination or socket wrench—a slotted screwdriver will not give enough leverage to seat the clamp securely). That's it. Attach your new hose to the faucet and celebrate your leak-free victory!

Easy Drip Irrigation for Every Garden

After a few years of watering with sprinklers or with a spray nozzle, you may want to make the leap into drip irrigation. This is an incredibly efficient way to get water to your plants. The times I've used a drip system, it has been amazing. I never had to think about watering; everything was set on a timer and went off on schedule. My garden was lush, plants grew to be gigantic, and my harvest was huge. But drip irrigation requires planning, basic plumbing skills, ordering parts, building and installing your system, as well as maintenance and diligence to keep it going.

Drip irrigation applies water directly to the soil through emitters or low-volume microsprayers.

WATERING YOUR LAWN

I don't water my lawn. My grass is in the backyard, where the soil is silty and holds water for a long time. Only in very hot summers does my lawn go completely dormant brown. The late spring and early fall rains in the Pacific Northwest green up my lawn nicely without irrigation. What if you want a green lawn?

Keeping a lawn green all year round takes a lot of water. Plan on giving your lawn about 1 inch of water per week during the summer; less during cooler weather. Do the tuna can test to determine how much water your sprinklers are applying. Place shallow cans or containers under your sprinkler spray area—at the edges and in the center. Turn on the sprinkler for 15 minutes. Measure the water in your cans and determine an average depth. Then do the math to figure out how long to run your sprinklers so that you give your lawn just 1 inch per week. If your sprinkler has a high output or puddles occur while you are watering, plan to water two or more times a week so that water soaks down to the roots, rather than running down the sidewalk and into storm drains.

FIGURING OUT HOW LONG TO WATER

AFTER 15 MINUTES OF WATERING WITH A SPRINKLER, IF YOUR CAN HAS:	INCHES	WATER FOR
	1/8"	2 hours
	1/4"	1 hour
	3/8"	45 minutes
	1/2"	30 minutes
	3/4"	23 minutes
	1"	15 minutes

They are efficient because they water just the soil, reducing evaporation and transmission of disease. You'll save time moving hoses and sprinklers. Fewer weeds will germinate, since only small areas of soil are moistened. Setting up a drip system is sophisticated and requires thorough planning. Your system can be customized to specific crops and set on a timer to run early while you are still asleep!

Soaker hoses are often used instead of drip tubing for a super-simple system that can easily be hooked up to a garden hose. Soaker hoses sweat water along the entire length of the hose, allowing it to seep out slowly over a long period. They can be moved around from season to season. Soaker hoses are available in ¼-inch and standard ½-inch sizes. They will not work efficiently in lengths longer than

50 feet or on steep slopes. Cover your soaker hose with mulch to keep it from degrading in sunlight. Run soaker hoses for 30 to 40 minutes per week. To test the length of time you need to water, run your soaker hose for 20 minutes and check the moisture an hour after you stop watering. Soaker hoses are made from recycled tires and there may be a concern about the safety and toxicity of these hoses (see sidebar, "Safety Issues in Watering").

Setting up a simple drip or soaker system

Start small and create a simple line for one or two veggie beds. Make a list of parts you will need and plan to go to the hardware store at least three times. If you are buying drip irrigation supplies online, order more than you think you will need and keep the receipt.

Keep your layout simple. Use in-line emitter tubing, which is much less of a hassle than individual emitters. The more emitters or tubing you have on each line, the more you have to look through to fix leaks or clogs—it's like finding the one faulty Christmas lightbulb in a massive holiday display. Cover drip tubing or soaker hoses with 2 inches of mulch to reduce evaporation. Use straw, leaves, or grass clippings for vegetable beds (woodchips work well for perennials or shrubs). See chapter 12 for suppliers and instruction books.

SAFETY ISSUES IN WATERING

Soaker hoses—those black hoses of varying lengths that allow water to seep slowly into the soil through tiny pinprick holes—are usually made from recycled rubber tires. Recycling rubber is a great way to reuse materials and keep useful items out of our landfills. But some gardeners are concerned that harmful chemicals may leach out of the recycled rubber into garden soil. If they do, are those chemicals taken up by the roots of edible plants?

There is little evidence to show that chemicals leach out of soaker hoses, though it's possible that's because few studies have been conducted on this. Evidence does exist confirming that ground-up tires made into a product marketed as rubber mulch—and sometimes dyed brown to

Basic Drip System Components

Backflow preventer or antisiphoning device: Keeps dirty water or fertilizers from being siphoned back into your home water supply (many new homes have hose bibs with built-in backflow preventers).

Pressure regulator: Delivers water evenly; prevents couplings and tubes from blowing out. Residential water pressure can be as high as 80 psi; most drip and soaker systems require 10 to 25 psi.

Filter: Keeps sediment from your home water supply from clogging the inside of your system.

Timer (optional): Inexpensive manual or battery-operated timers are an excellent choice for busy city farmers. Manual versions are like egg timers and require you to turn on the water and twist the dial to the length of time you need. Battery-operated timers can be programmed to water for a set length of time each day or a few times a week.

Half-inch solid mainline tubing: This attaches to your hose bib or garden hose and delivers water to beds. If this tubing bends, it will kink and stop the flow of water. Elbow- and T-shaped fittings are required to go around corners.

Quarter-inch in-line tubing: Emitters are preinstalled inside the tubing and are spaced at regular intervals. You can choose in-line tubing with emitters spaced 6, 9, and 12 inches apart, depending on your soil type and planting layout.

Quarter-inch solid tubing: Use this tubing to connect to microsprayers or span paths where you don't want to water.

Microsprayers: Tiny, low-volume spray heads are great for keeping seedbeds moist and for watering baby greens or fruit trees. The spray pattern can be adjusted to fit the shape of the bed. Microsprayers are fragile and easily damaged.

Various couplings, connectors, and adaptors: There are a number of hose couplings, tubing connectors, and other gizmos that are required to build a more complicated system.

---

look like woodchips!—leach contaminants into the soil. Those contaminants then filter into our water systems, finding their way to streams and rivers where they affect fish and other wildlife. But there is no direct evidence that this happens with soaker hoses, which have much less surface area than the small pieces of rubber mulch.

Other hoses in your garden may pose potential health risks too. You might notice a warning on the garden hose you buy: Not safe for drinking. This refers to the inner PVC coating and brass fittings, which contain lead. Studies show that lead leaks out, making it an unwise choice for a quick drink while you are out watering the garden.

But what about the garden? Lead is fairly immobile in soil, because it gets bound by other elements; it is not taken up by roots and is not washed away. Concerned gardeners can find lead-free hoses marketed for marine or RV uses and new reinforced garden hoses made from medical-grade plastic resin with nickel-plated brass fittings. Even so, it's always a good idea to flush out the hose before taking a sip.

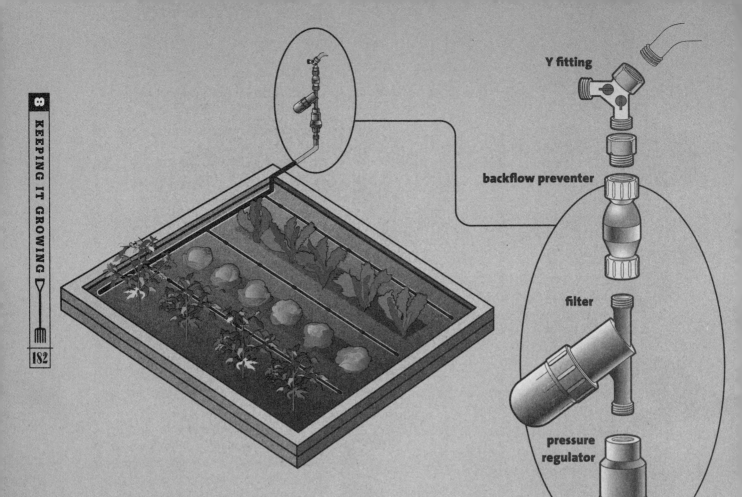

Y fitting

backflow preventer

filter

pressure
regulator

garden
hose fittings

Building the Main Assembly

Build your main assembly and attach it to a Y fitting at your outside faucet. There are three essential parts that should be installed in every drip or soaker irrigation system.

STARTING AT YOUR HOSE BIB:
1. Backflow preventer or antisiphoning device
2. Filter
3. Pressure regulator
4. Timer (optional)

If you decide to use a timer, it should be placed after the pressure regulator. Attach your garden hose to the end of the main assembly, then to the half-inch main line at your garden beds.

KEEP TUBING IN ITS PLACE

Your drip tubing and soaker hoses have a mind of their own and will twist and curl uncontrollably unless you anchor them somehow. Bricks and rocks can work in a pinch or you can buy wire soil staples to hold down your soaker hoses or drip lines. Resourceful city farmers make their own staples. Use 12 inches of heavy gauge wire (could be old coat hanger) and bend it into a U shape that is wide enough to span your hose. Make sure your staples are at least 4–6 inches long.

Winterizing

Protect your drip system from damage caused by freezing temperatures. Remove and bring the main assembly and timer inside during the winter—these parts cannot tolerate a hard freeze. Flush the main lines by removing the end caps and running water through the lines, washing out any sediment. Then empty all lines of water. If you live where it doesn't freeze in the winter, you can leave your drip lines in the garden. Coil up quarter-inch tubing and tie the loops. Coils of tubing will be easier to find in the spring. If you live where it freezes for several months, bring your drip lines inside so that residual water in tubing doesn't freeze and damage emitters or microsprayers.

Harvest the Rain

One way to conserve water is by collecting it in a rain barrel. Simply put a rain barrel under a downspout and use this water for your ornamental (nonedible) plantings. This water may not be safe for edible crops because of the heavy metals in most roofing materials (see sidebar, "Roofing Materials and Water").

There are many sources for ready-made rain barrels and custom bladders that can be hidden under a deck. Resourceful

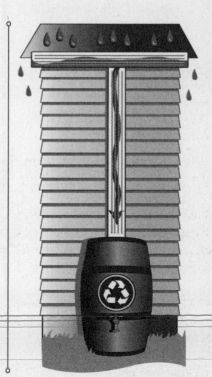

ANATOMY OF A
RAIN BARREL

Downspout fittings: Add an elbow or short sections of downspout so that rainwater goes directly into the barrel. You may want to invest in a downspout adapter that will allow you to direct water into the barrel when you need it or divert to your regular drainage area when your barrel is full.

Top screen: A screened lid where the downspout will drain. The screened lid should keep mosquitoes from entering your rain barrel to lay eggs. Keep spigots capped or hoses attached to prevent mosquitoes from crawling into the fittings.

Bottom spigot: Near the bottom of your barrel is a faucet for filling your watering can or attaching your garden hose.

Overflow spigot near the top of the barrel: Attach a hose to this spigot so that excess rainwater can be directed where you want it. Link several barrels together by connecting the top spigots with a short piece of garden hose. Make sure that overflow water is funneled at least 8 feet from the foundation of your home to avoid basement flooding.

city farmers can obtain recycled 55-gallon drums from local food processors and add their own spigots. Recycled barrels should be food-grade quality—many carried olives, peppers, or pickles before being reinvented as water reservoirs.

Setting up your rain barrel

Place your rain barrel under the gutter downspout that gives you easy access and is close to ornamental beds (if possible). Set the barrel on a 2-foot-tall cinder-block stand to increase water pressure and make it easier to get your watering can under the spigot. Rain barrels generally only hold 55 gallons. Linking several barrels together will increase storage capacity.

Maintaining your rain barrel

Rain barrels require little maintenance. Keep your gutters clear so that rainwater flows unobstructed into your barrel. Keep the top screen clear to prevent clogging and make sure the overflow spigot is clear. After several years, empty the barrel, open the top, and wash out the inside with mild dishwashing soap.

Roofing Materials and Water

Not all roofing materials are created equal. Before investing in a rain barrel, make sure you can use the water you collect.

Enameled steel and glazed tile roofs leach little or no contamination. Rainwater collected from these can be used on vegetables and other edibles.

Asphalt shingle roofs contain various compounds that are toxic if ingested, but water collected from this type of roof is fine to use on shrubs, trees, and other plants you don't intend to eat.

Wood shingles or shakes that have been treated with any chemicals are toxic and water should not be collected in a rain barrel from roofs made of those materials.

Copper roofs or gutters may leach heavy metals into rainwater, so this water should not be used in your edible garden. Don't collect rainwater if you have a zinc antimoss strip along the top of your roof.

Profile:
WINDOWFARMING

WANT TO GROW FOOD AND DON'T HAVE A YARD?
All you need is a sunny window in your own apartment or at the office. The Windowfarms Project shows you how to grow up to 25 plants in a typical 4 x 6 foot window.

The group's Web site defines the system as "a DIY (do-it-yourself) vertical hydroponic farming system for urban dweller's windows." Artist Britta Riley began Windowfarming in early 2009 and still works on the project.

Plants are grown in small pots set in recycled plastic water bottles; the bottles are arranged in columns that hang in your window. Water and nutrients are pumped up through the pots. The Windowfarms Project offers various sizes of kits for sale and provides tons of information on how to set up and run your farm. The community of users shares stories and tips online.

Window farms work not only for apartment dwellers who have no garden space outdoors, but also for those who live in cold climates where no winter gardening takes place.

What will you grow in your window farm? Greens for salads are the most popular choice, but you could also grow peas, radishes, cherry tomatoes, and herbs. Think of harvesting enough for a fresh salad each week in the middle of winter!

LEARN MORE:
windowfarms.
org

Watering Your Thirsty Plants

Watering your city farm—whether it's a windowfarm, a container garden, or a garden plot in a sunny spot on your property—is one of your most important jobs as an urban farmer. Set up a simple system that will work for you, using quality parts that will last for years. Experiment with ingenious watering devices that conserve water and bring moisture right to the plants. Your plants will thank you!

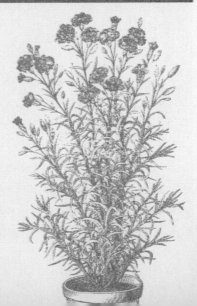

LOVING YOUR ENEMIES

EST OR DISEASE INFESTATIONS often reveal that a plant is not getting what it needs to thrive. Most pest or disease problems can be fixed simply by building healthy soil, by making sure the plant is in the right place and is getting adequate water, and by planting a diverse garden to encourage beneficial creatures to take up residence there.

If that doesn't work, we propose the organic approach to pest, weed, and disease control. This means, first, understanding the underlying conditions that make it favorable for the pest or disease to thrive, then changing things so that either the plant gains vigor and outgrows the problem or you encourage natural predators to maintain the balance and take care of things for you.

Organic Pest Management for Vegetable Gardens

We suggest taking the organic approach to dealing with pests and diseases on your city farm. Below is the sequence for preventing, assessing, and handling any curiosity in the garden.

Prevention

Good gardening practices go a long way toward preventing disease or pest infestations. Keep your plants healthy by building healthy soil, rotating crops, putting the right plant in the right place, adequately watering and fertilizing, and properly pruning or training your plants to promote air circulation. Planting a diverse garden provides habitat for beneficial insects and birds. Practicing good sanitation by removing infected plant material will also help stop the spread of diseases or harmful insects. Diseased plant material can be given to livestock or put in your garbage or curbside yard-waste recycling receptacles.

Positively identify the troublemaker

You can't take action against any pest or disease until you understand what is causing the problem or if there really is a problem at all. You must positively identify the creature and learn about its habits and habitats before you can take any steps to eradicate it. Over 95 percent of the creatures in your garden are not harmful to you or your plants. Before you cast blame on some crawling bug, make sure it really is a pest. Ladybug larvae are pretty fierce looking—and when they are present in large numbers, that can be worrisome. In reality, however, these larvae are just eating aphids, insect eggs, and mealy bugs.

After you observe damage to your plants, such as holes in the leaves, look for the culprit. Notice bite marks or patterns on leaves and insect frass (poop). Many plant-eating larvae or caterpillars are nocturnal and feed at night. Go out with a flashlight after dark and look for the creature that is eating your plants. If you notice disease symptoms, such as powdery mildew on leaves, check daily to see if that symptom is spreading on the plant or moving to neighboring plants. Use books and other horticultural resources, plus the Internet, to positively identify pests or diseases afflicting your garden.

Learn habits and habitats

After you have identified the creature or disease blighting your garden, learn about its life cycle, feeding habits, and natural enemies or controls. Some pests only do leaf damage in the larval stage, such as leaf miners on Swiss chard. It is easy to remove leaves and increase plant vigor with liquid fertilizer so the plants can outgrow any damage.

Beneficial insect larvae often look very different from their mothers. Harmful insects are food for these young garden heroes. Being able to identify the different stages of insects helps you to figure out who the good guys are!

Practice tolerance

There are diseases, ravenous larvae, and weeds in every garden—it would be weird and wholly unnatural if there were not. After you know about the creature and how it affects your garden, decide if this is something you can live with or if you really need to get rid of it.

Use the least toxic approach, and see what happens

In time, nature will sort it out for you. Since pests are an important food source, waiting and watching may be the best approach. The aphids on your roses may look awful, but if you have an active insect and bird community in your garden, this is like a feeding station.

Aphids are the plankton of the garden—they are food for many different beneficial insects and birds. Just watch: Ladybugs and lacewings will show up along with small songbirds, yellow jackets, hornets, syrphid flies, and tiny parasitizing wasps. These creatures and their babies are all there to grab a juicy, sweet aphid for lunch! If your aphid infestation reaches the gross point, reduce the population by hosing them off with water or smashing them with your fingers.

Physical controls

Direct controls are ways to physically block or remove pests or diseases. This includes pruning diseased leaves, hand-picking bugs, and using physical barriers, such as a floating row cover or copper barriers.

Biological controls

By introducing the natural predators of your pests, you can biologically control infestations. It's a wild kingdom out there— with all creatures eating and being eaten. Letting the bugs sort is out (or introducing additional predators) is extremely effective in managing insect pests, though this may take time.

WHAT IF THINGS GET OUT OF HAND?

What if dandelions take over the yard? Figure out what it will take to fix the problem and then decide if it is worth it. There are almost always several solutions to a problem. Dandelions all over the lawn could call for a complete turf renovation or could be loosened and removed by hand as a Zen meditation. You could keep your lawn nicely mowed to reduce flowers or collect flowers to make a farmstead classic—dandelion wine.

Last resort: Remove the plants

I don't spray any toxic chemicals on my lawn or in my home, so I never even think about devising poisonous solutions for garden problems. I figure I encounter lots of icky stuff just driving a car, working in an office, and living in a city—why intentionally add poison to my garden environment? If I have tried everything and the plant is still infected, I take it as a sign that this isn't the time or place for that plant. Rather than squirting poison into my environment, I just remove the plant or move it to another location.

Bugs Control Weeds, Too!

Some insects control invasive weeds. Most plant-eating insects consume only one kind of plant their whole lives. One summer we had a weevil infestation on the mullein in the Children's Garden. We learned that the mullein weevil only eats mullein and lays its eggs in the seedpods that form on the tall flower stalks. Each pod makes a quarter-million tiny seeds that are dispersed as the wind and animals move the stalks about. The weevil larvae hatch and eat the immature seeds in the pod before they pupate. These weevils are used in agricultural fields to manage mullein—we haven't had a bumper crop since the summer of the weevil!

Common Beneficial Creatures of the City Farm

When you plant a garden, bugs will come—and that's a good thing. It is reassuring to know that most of the insects, spiders, and other crawling (and flying) creatures on your city farm are either helpful or provide food for beneficial organisms. Good bugs help the garden by eating other creatures to keep balance or by pollinating flowers. Invite beneficial insects by leaving some areas wild and unmowed, building bug houses, spreading mulch, and incorporating a diversity of flowering plants in your garden.

Get to know some bugs

Learning about the insect and spider community in your garden is a surprisingly fun aspect of city farming. My bug education has been influenced by my work in the Children's Garden. For 15 years I have spent most days exploring the garden with people who are less than 4 feet tall. Children are close to the ground and notice things that we might pass by unawares. With close observation and some handy reference guides, you can learn a great deal about garden critters, which include anthropods (arachnids, myriapods, and isopods) and other invertebrates like worms and slugs.

MIGHTY ARTHROPODS

Many of the crawling and flying creatures that you'll find on your city farm belong to an amazing group of creatures that share the same traits—arthropods. Among the characteristics of arthropods are a hard exoskeleton, bilateral symmetry, jointed legs, a segmented body, and many pairs of limbs. They must shed their exoskeleton or molt in order to grow. Besides insects, arthropods on your city farm include spiders, centipedes, millipedes, and sow and pill bugs.

Spiders (arachnids) are a major garden predator and eat other arthropods. All spiders have two fused body parts: Their abdomen is connected to the head and thorax, called the cephalothorax, where legs, mouth, pedipalps, and eyes are attached. Most spiders have poor eyesight and use their sense of touch to hunt prey. They catch prey with or without a web. Webs can be orb, funnel, or tangled in shape. Babies look like miniature adults; eggs are laid in cottony egg sacks that are attached to stationary objects or carried around on the spider's abdomen.

Centipedes and millipedes (myriapods) are arthropods with lots of legs attached to a long, slender, segmented body. Both inhabit mulch and organic debris. Centipedes are predators and feast on smaller bugs and slug eggs. They are brick-red, quick-moving, and have big head and jaws. Their bodies are flat, with one pair of legs sticking out of the side of each segment. Millipedes only eat dead stuff—plants and animals. They snake slowly around the compost pile on a gazillion tiny legs. They curl up into a spiral when disturbed.

There are two different arthropods, called isopods, common on all city farms. Pill bugs and sow bugs (isopods) are crustaceans—the terrestrial cousins of crabs and ancient trilobites. They breathe with gills and have a segmented, armored body with seven pairs of legs. Pill bugs, or roly poly bugs, resemble an armadillo and roll up. Sow bugs, or potato bugs, are flatter, have a fringed edge, and don't roll up. They are good bugs and primarily eat dead plants. These creatures love to live in the compost or worm bin and consume wood debris and other organic matter.

Slimy invertebrates live in most healthy garden ecosystems. Earthworms, red wrigglers, and threadlike pot worms are helpful residents since they aerate garden soil and make rich compost. Most other wormy things you find are insect larvae. Slugs and snails are uninvited guests of the city farm. They eat decaying organic matter and provide food for beetles and centipedes, but they also love your delicious vegetables! Slugs and snails can hide, but they can't run. They are easy to pick off plants and relocate to weedy areas or feed to the chickens or ducks.

Insects are the most numerous of the small creatures that you will find on your city farm. Ladybugs, beetles, flies, butterflies, dragonflies, bees, and ants are just some of the insects that live in your garden. All insects have two compound eyes, a pair of antennae, and three body parts: Head, thorax, and abdomen. All insects have six legs and most have two pairs of wings. Insects live a short time, typically one year. Some insects eat plants, while others eat insects.

Insects are encased in a hard exoskeleton, which they must shed in order to grow larger—their new, bigger exoskeleton is folded up underneath. Each time an insect sheds its exoskeleton it is called an instar. Adults lay eggs. Insect larvae (such as caterpillars) must pupate to become adults (butterflies or moths). Some insect babies, called nymphs (such as spittlebugs) gradually become adults and develop wings as they molt.

We love insects in the Children's Garden, but they have a lot of body parts to remember. We use this great song from our friends at Tickle Tune Typhoon and body movements to act out the parts of an insect and how everything fits together.

I Am an Insect in This Life

Lyrics by Dennis Westfall, recorded on *Tickle Tune Typhoon's Singing Science*

(To the tune of "Bingo")

C   F   C
I am an insect in this life
C   G7   C
And this is what I look like:
C   F
Two antennae, compound eyes,
G7   C
One, two, three, four, five, six legs
C   F
Two sets of wings so I can fly,
    G7   C
My thorax and my abdomen.
    G7   C
And this is what I look like.

Bridge
(To the tune of "Old MacDonald")

C   F   C
Instead of bones my body grows
    G7 C
An exoskeleton
                    F   C
It's hard and strong and seals me in
    G7   C
Protecting my organs

My Heroes!
Now that you know what to look for,
here is a list of some common beneficial
insects that you may find on your
city farm.

LADYBIRD BEETLES OR LADYBUGS

(COCCINELLIDAE)

- The most recognizable insect in the garden. Subject of songs and wishing games, this cute bug is really a bloodthirsty killer.

- There are 250 different species of ladybird beetles. They vary in size (1/8"–3/8") and come in yellow, orange, red, gray, and black—with and without spots.

- Larvae are alligator-shaped, black with orange splotches. They look like a bug that you might need to worry about. Adults and larvae are ravenous; they eat aphids and other insects.

- Bright yellow or orange eggs, shaped like a football, are laid in clusters near aphid colonies, under leaves or on twigs. Larvae hatch in 3 to 7 days.

- Ladybugs typically have 5 instars before they pupate. Look for discarded exoskeletons where the larvae are feeding.

- The pupa is attached to leaves, sticks, or garden sheds—it is a black-and-orange shell that almost looks like a ladybug. The adult emerges from one end.

- Adults lay eggs throughout the season. The whole life cycle can occur in 3–13 days, depending on weather and food.

- Adults overwinter in fallen logs, the siding of wooden buildings, or rock gardens.

- They primarily eat aphids, also mealy bugs, scale, and other soft-bodied insects

- Adult lacewings, which are bright green insects, are 1/2-inch long with two pairs of large transparent wings. Adults eat aphid honeydew or nectar and fly around looking for a mate.

- Lacewing larvae, nicknamed "aphid lions," are ravenous and eat aphids, mites, and small insect eggs. Larvae measure 1/3-inch at maturity and resemble alligators with dark- and light-brown markings.

- Pale green or gray eggs are laid on top of threadlike shafts on plant leaves in the middle of aphid colonies.

- Lacewing larvae use their curved mandibles to skewer aphids and suck their vital fluids.

- Three to four generations are produced per year.

- Lacewings spend the winter in the pupal stage.

LADYBUG, LADYBUG, FLY AWAY HOME!

It is exciting to buy a pack of chilled ladybugs at your garden store to release in the garden. Ladybugs sold as predators are harvested when they are hibernating. They have lots of fat energy stored up so that they can fly away from their wintering grounds to find food. Those that you purchase are programmed to fly, fly away. Releasing purchased ladybugs in a neighbor's yard might be your best bet, but there's no telling which way this red beetle will fly. Save your money and leave the aphids alone—predators will eventually show up.

GREEN LACEWINGS
(CHRYSOPIDAE)

GROUND BEETLES

(CARADIBAE)

- This common, ground-dwelling garden insect looks beautiful—black with purple iridescence—but it's a voracious predator. There are many different ground beetles that are skillful predators in your garden. They are ferocious-looking, run very quickly, and have large mandibles.

- Larvae have long, segmented bodies with six legs attached near the head. Whitish gray with red head and thorax, they are quick-moving and have noticeably large pinching mandibles.

- Larvae eat insect eggs and other ground-dwelling insects. Larvae pupate in the soil without a cocoon or shell (it's creepy to find an albino beetle pupa in midtransformation when you're transplanting).

- Eggs are laid in the soil, and there's one generation per year. Adults hibernate under mulch and plant debris.

- Adults eat cutworms, root maggots, and slug eggs.

- They live in moist soil under leaves, in a compost pile, or under rocks or burlap sacks.

- There are 3,000 different species of rove beetle in North America.

- These large (1/2"–1") black beetles are ferocious-looking and put on a menacing show—curving their abdomen and opening their huge mandibles, resembling a scorpion—when they feel threatened.

- They are fast-moving with an elongated, slender, segmented body.

- Rove beetles live in garden debris, under rocks or bricks, and are most often found in decaying matter and carrion.

- Rove beetles eat cutworms, slug eggs, and other invertebrates.

- Their eggs are laid in soil or mulch.

- They may overwinter as larvae, pupae, or adults, and there are several generations per season.

- The rove beetle larva resembles the ground beetle larva—a cream-colored, elongated body with a large reddish head and mandibles.

ROVE BEETLES

(STAPHYLINIDAE)

CENTIPEDES

(CHILOPODA)

- This big, red, fast-moving myriapod is a fierce predator ($1/2$"–$1^1/2$") and is at the top of the food chain. You may also find a long, yellow, threadlike relative in your garden.

- Centipedes have flat, segmented bodies with one pair of legs per segment that stick out to the sides. Smaller, slow-moving millipedes have two pairs of legs per rounded segment, with legs pointing down toward the ground.

- Each centipede has a large head with pinching mandibles that grab prey—the centipede's jaws are too weak to pierce our skin.

- They lay eggs in soil, decaying yard debris, or worm bin bedding.

- Baby centipedes look like miniature adults; they produce more than one generation per year.

- Centipedes feed on slugs and slug eggs, worms, fly larvae, and some spiders.

- They live in moist, decaying matter and under rocks in the garden. They are commonly found in compost and worm bins.

- Add mulch and organic matter to garden beds to encourage centipedes.

Insect Munching

If you wonder whether a bug you come across
in the garden should stay or go—look at how the mouth is
formed. Insects (and spiders) with pincher-type mouths do not
eat plants—these mouths are built to eat other bugs.
Bugs with mouths that chew or suck are plant eaters—
hand-pick those bugs and feed
them to the chickens.

Syrphid or hover flies
(SYRPHIDAE)

- The adult has black and yellow stripes and looks like a small bee (1/3"–1/2") that darts quickly about and hovers above flowers.

- Adults each have one pair of triangular wings and gigantic compound eyes on a large head.

- Larvae are blind and look like light green slugs. They blend in with green plants and aphids. They leave a trail of tarry frass (poop).

- Larvae feed on aphids, mealy bugs, and other soft-bodied insects (like caterpillars and maggots) throughout the season.

- White, oval eggs are laid singly on foliage.

- They spend winter in the pupal stage.

DRAGONFLIES & DAMSELFLIES

(AESCHNIDAE)

- Dragonflies belong to a small family of aquatic insects that are voracious predators. Fossils of giant dragonflies date back 250 million years.

- Finely netted wings characterize both dragonflies and damselflies. Both have a long thin abdomen with a sturdy thorax and large compound eyes.

- Dragonflies are larger and sturdier (³/₄"–4") and have wings that lay outstretched, horizontal to the body. Damselflies are dainty (³/₄"–1³/₄") and hold their wings vertically above their body.

- Nymphs live and transform in the water—they look nothing like the adult.

- Adults and nymphs eat insects— especially mosquito larvae.

- To encourage dragon and damselflies, provide a water source, such as a pond or a large birdbath.

COMMON GARDEN SPIDER

(ARANEUS QUADRATUS)

- The common garden spider found around your city garden or farm ranges in size from ¹/₈" to 1 ¹/₄" inch. An important predator, the spider keeps everything in balance.

- It spins an orb web with spokes radiating from the center to catch flying or crawling insects. It spins a new web each day, recycling old silk from torn webs.

- All spiders have 8 legs, 2 fused body parts, a pair of pedipalps, a piercing jaw, and 8 eyes. Most have poor eyesight and depend on the hairs on their legs and body to help them hunt.

- They wrap insects caught in their webs in silk, inject some poison into them, and then suck up a delicious insect smoothie!

- Spiders live about a year. The new generation overwinters in a cottony egg sack.

- Spiders feed on insects year-round. They are shy and quick-moving if surprised.

CRAB

SPIDERS

(THOMISIDAE)

- There are many different species of crab spiders that live on city farms. They are important predators that eat a variety of insects and help keep the garden ecosystem balanced.

- They don't spin webs but hide in the petals of flowers, waiting to grab an unsuspecting insect.

- They range from 1/8 inch to 3/4 inch in size and vary in color from brown to green. Some species can change color to match foliage.

- This species carries the name crab spider because the front three sets of legs are pointed forward, resembling its aquatic namesake.

- They have terrible eyesight and rely on their sense of touch for hunting—they hunt day or night.

[Recipe] Kids and Bugs: Bug Houses Make Exploration Fun

Building bug houses and exploring the creature world is a great way for big and little people to learn about everything that helps make a garden grow. Caring for creatures makes sense to kids and helps them learn to care for all life on the planet. Creating shelter for ground-dwelling beneficial insects and spiders will help maintain biodiversity in your garden.

WHAT YOU WILL NEED

Rocks and bricks

Sticks and broken pottery

Hand trowel or small digging implement

Old tofu or yogurt tub

Water

Plants for the bugs to eat

Sidewalk chalk, fabric, art supplies

Magnifying box or lens

Bug books and field guides

1. Find a place in the garden that is partly shady and where you can keep the ground moist. The end of a garden bed works nicely.

2. Dig out a shallow hole as a water reservoir (or swimming pool), put in the tofu or yogurt tub, and fill the tub with water.

3. Gather plants and flowers for the bugs to eat and place these artfully here and there around the water.

4. Place rocks and bricks on the soil around and over the water reservoir. Add sticks, broken pottery and leaves as beds, doors, slides, helicopter landing pads, and other fun bug house furnishings.

5. Decorate your bug house with flowers, draw on the rocks with sidewalk chalk, and hang signs around your dwelling.

6. Keep the bug house area moist to encourage creatures to "move in" and leave everything undisturbed for a couple of weeks.

7. Explore! Carefully take your bug house apart, explore and identify the creature life, and then rebuild!

After your bug house is built, it is very hard to keep from peeking to see if something has moved in. Talk with your child about habitat destruction and disturbance—if rocks are continually moved around, no bug can make the new house its home. Pass the time between building and exploring by reading bug books together and exploring the creatures of the air.

OR TRY THIS INSTEAD

If bugs are too icky or scary for you or your child, think about building small fairy houses. These can be fanciful creations with sparkly decorations and colorful flags that attract magical fairy folk to live in your garden. Check out some books from the library about fairies and think small. Have fun!

Mason bees
(OSMIA LIGNARIA)

- This cute blue-black bee is an important early pollinator of fruit trees—apples, pears, cherries, and plums.

- This small, docile insect (3/8"–5/8") lives only a short time (mid-March to June), yet visits over 200 flowers each day to collect pollen for its young.

- Adults deposit a pile of pollen in holes found in old tree snags or in mason bee homes. Then they lay their eggs on this pile of food. They seal up the chamber with mud (hence the name mason bee) and start building another pollen pile in the tunnel. Each tunnel may have 5 nursery chambers.

- The egg hatches and the larvae eat the pollen and grow during summer and fall. In the winter they pupate in nursery chambers; in the spring they emerge from the tunnel as adults.

- Encourage mason bees by hanging bee homes on the south side of your house under the eaves—sheltered from rain and strong winds. Hang bee homes before fruit trees bloom. Make your own bee houses by drilling 3-inch-deep, 3/8-inch-diameter holes in old logs or two-by-fours.

Honeybees
(APIS MELLIFERA)

- This is your classic bee—fuzzy with muted yellow and black stripes on its abdomen. It is about 1/2– 2/3 inch in size. The most important pollinator of fruits, vegetables, and flowers on your city farm.

- They are more interested in nectar than in you; they are not aggressive toward humans unless threatened. Using their stinger is a suicide mission, since they die after stinging.

- Worker bees feed on nectar and carry pollen from plant to plant. Look for the bright yellow saddlebags of pollen that honeybees carry around.

Bumblebees

(BOMBUS SPP.)

■ Eggs are laid continuously, except during extremely cold temperatures. Adults hibernate in the hive.

■ Honeybees nest in fallen logs or snags and in bee boxes.

■ They go where the flowers are, so encourage an active bee community by planting a variety of flowers and flowering herbs, such as rosemary, oregano, sage, and lavender.

■ These fat, jolly bees are a favorite in the garden. Like great lumbering cargo planes, ranging in size from 1/4 inch to 1 1/4 inches, they buzz steadily around the garden, sipping nectar and moving pollen.

■ There are many different species found on city farms. Some are truly huge, while others are more petite. Some are completely black; others sport the classic yellow-and-black stripes. Observe carefully: How many different types are visiting your garden?

■ They nest and lay eggs in the ground; their larvae eat pollen and nectar. They produce several broods a year.

■ Each colony lasts only one year and produces a fertilized queen that overwinters in the ground. She emerges in spring to establish the next colony in a new spot. The other bumblers die in the winter.

■ Encourage them by leaving an unmowed, weedy area as their habitat, and planting a wide variety of flowering plants.

PARASITIC WASPS

(HYMENOPTERA, SPP.)

- This is a group of tiny, seldom-noticed predatory wasps that don't bite or sting people. You will not see these important insects, but you may notice the results of their work—mummified aphids or slain caterpillars.

- There are quite a few parasitic wasps—such as the Aphid wasp, Trichogramma, Braconid, Encarsia, and Ichneumon—busy working in your garden. Different species prey on aphids, whiteflies, cabbageworms, and hornworms.

- They reproduce by laying their eggs on or inside a pest host.

- Aphid wasps, for example, deposit an egg in the aphid. The larvae hatch inside and eat the aphid's internal tissue and organs. The mummified aphids, bloated and brown, are quite distinct from living aphids. Inside the mummy, the tiny wasp spins a cocoon, pupates, and emerges as an adult through a hole in the top of the aphid.

- Encourage parasitic wasps by planting a variety of flowering plants and vegetables and leaving nature alone!

Wasps, yellow jackets and Hornets

(VESPULA SPP.)

- It's hard to view these annoying, uninvited picnic guests as fierce warriors and important predators.

- Social wasps (this includes hornets and yellow jackets) start a new colony each year. The fertile or founding queen begins to build a nest out of wood fiber in trees, under roof eaves or in the ground.

- An egg is laid in each papery cell. The hungry larvae are blind maggots that eat insects. The founding queen collects insects from the garden to nurse her young; when they are adults, they will do the hunting while she focuses on laying more eggs.

- Adults capture insects, such as flies, thrips, aphids, and crane flies, to feed their hungry young. Larvae give the adults a drop of nectar that is the energy they need to live.

- Larvae pupate in the nest. Many pupate prematurely to serve the colony as nursemaids.

- Each colony produces 20–50 fertile queens that will overwinter in mulch or garden debris (or the woodpile) and emerge in the spring to start the cycle once more. Fertile queens are easy to identify: They are larger than normal and appear in early spring and late fall—they are loaded with eggs.

FLOWERS TO ATTRACT BENEFICIAL INSECTS

Plant flowers to attract beneficial insects. Invite all creatures into your garden and let them keep the balance. Beneficial insects, in particular, love certain flowers.

FLOWER	ATTRACTS
Anise	Parasitic wasps, tachinid flies, lady beetles
Anjelica	Lacewings, lady beetles, parasitic wasps
Annual candytuft	Hoverflies SHELTERS ground beetles
Buckwheat	Hoverflies, braconid wasps
Cabbage Tribe	Lacewings, lady beetles
Dill or Fennel	Lady beetles, wasps, spiders, hoverflies, bees, yellow jackets, hornets
Lovage	Beneficial wasps SHELTERS ground beetles
Marigolds	Hoverflies, parasitic wasps
Monarda	Bees, parasitic wasps, beneficial flies
Nasturtium	Aphids SHELTERS ground beetles and spiders
Scabiosa	Hoverflies, tachinid flies
Sunflower or *Asteraceae* family	Hoverflies, lacewings, parasitic wasps, lady beetles, tachinid flies, bees, wasps
Zinnia	Lady beetles, parasitic wasps, parasitic flies, bees

THE BAD GUYS

Most insect pests, while troublesome for the city farmer, are part of a big, complex ecosystem. Without them there would be no food for the good guys. Learning about their habits and habitats can help you keep populations in check until the beneficial creatures arrive.

THE BAD GUYS

APHIDS

(HOMOPTERA FAMILY—MYZUS PERSICAE)

- These piercing, sucking insects attack the watery, soft tissue in plants, especially new growth. They produce sticky honeydew.

- Aphids undergo countless life cycles in a season—they seem to be constantly reproducing!

- They attack juicy plants and those that are stressed because of poor soil or inconsistent water. Nasturtiums and the cabbage tribe are among their favorites.

- Control aphid populations by squishing or spraying them off the plant with a heavy stream of water.

- Predators include ladybugs, dragonflies, wasps, birds, and spiders. To encourage these beneficial insects, plant flowers such as lavender and feverfew.

- Homemade soap spray is an effective control since it dries out and kills these soft-bodied insects. Just add 2 or 3 drops of dish soap to a spray bottle filled with water. Use this spray on both sides of leaves or stems to smother the aphids. Apply to a small area of infected plant first to make sure soap doesn't harm the plant.

- Ants love aphid honeydew and will "farm" a colony for this sweet sugary goodness. Ants will fight to keep other beneficials away. Disrupt the ants by keeping soil moist around the plant and reducing aphid population with a strong blast of water.

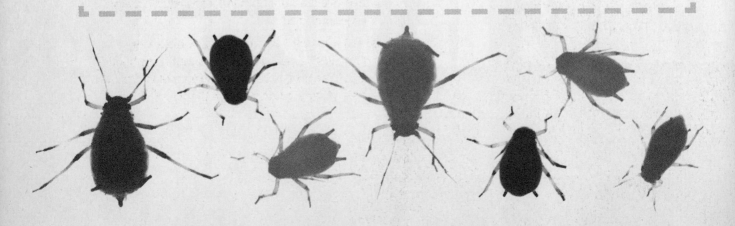

IMPORTED
CABBAGEWORMS/
CABBAGE BUTTERFLIES
(PIERIS RAPAE)

- These are white butterflies with one brown dot on each wing. Caterpillars are slow-moving, and have velvety-green skin with pale yellow side stripes. Eggs are pale yellow to orange, shaped like a football standing on end. Butterflies lay eggs singly on brassica leaves.

- There are two to three generations per year. Larvae pupate in the soil—adults emerge as the soil warms up in early spring.

- They feed on the cabbage tribe—broccoli, cauliflower, kale, kohlrabi, collard greens, arugula, mustard greens, radishes, and turnips.

- They do the most damage in the larval stage, eating holes in the leaves of your plants and leaving behind piles of green frass (poop). These green caterpillars can be found during the day on the leaves.

- Control by hand-picking the caterpillars or covering the target crop with a floating row cover to prevent adults from landing and laying eggs on your leaves.

- Biological controls include parasitic wasps, tachinid flies, lacewings, birds, and some beetles. If you see dead caterpillars on your plants, that is a sign that predators are on the scene!

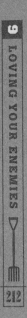

CARROT RUST FLIES

(PSILA ROSAE)

MAGGOTS AND DAMAGE ON VEGETABLE

MAGGOT

ADULT

■ The adult is a small black or green fly with yellow hair, head, and legs. Adults are hard to identify. Eggs are laid in the crowns of plants in the carrot family. The light yellow larvae hatch and tunnel through the root, leaving soft dark orange frass (poop) that rots quickly.

■ Two or three generations are produced each year. Larvae and pupae hibernate in the soil.

■ Adults lay eggs on carrots, celery, parsley, and parsnips.

■ A floating row cover is a simple and effective barrier. After carrots have germinated and are a couple inches tall, cover the bed with a floating row cover. Remove the cover when you water.

- A fat, greasy-looking caterpillar found wedged in between leaves or around the base of the plant (in the top inch of soil). The larvae of nondescript moths, they have a black head with a big chewing mouth, and curl up to form a C when disturbed. They range in color from brown/gray to green, depending on their diet.

- There are up to five generations in a season. Larvae are seldom seen during the day. Eggs are laid in the soil.

- Larvae pupate in the soil. Pupae are chestnut-colored, segmented capsules that are the diameter of a pen and about an inch long.

- Larvae eat the stems of young vegetable and flower seedlings at the soil level.

- Put a cardboard collar around seedlings (or seeds when you plant) as an effective direct barrier. Cut a toilet paper tube about 2 to 3 inches long. Push tube into the soil so that about an inch is above the ground. Remove the tube after the seedlings are 12 inches tall. Dispose of the pupae when you find them.

- Trichogramma wasps are a natural predator.

CUTWORMS

(NOCTUIDAE SPP.)

■ This is the larval stage of several species of small flies. Flies lay their eggs on the underside of host plant leaves, and, as the eggs hatch, the larvae tunnel into the leaf and begin feeding on the inner membrane.

■ There are several generations per year. They spend the winter as pupae in the soil.

■ The damage they cause looks like brown windows or mazes on leaves—if held up to the light, you will see little worms inside.

■ Primarily found on spinach, beet greens, and Swiss chard. You may also see leaf miner trails in beans, blackberries, cabbage, lettuce, peppers, potatoes, and turnips.

■ Direct controls include removing damaged leaves, removing eggs from leaves, and covering the crop with a floating row cover to prevent the flies from laying their eggs. Increase the vigor of the infested plant with an application of liquid fertilizer and it will outgrow any damage. Remove the damaged part of the leaf—the rest is fine to eat.

Leaf MINERS

(LIRIOMYZA SPP.)

CODDLING MOTHS

(CARPOCAPSA POMONELLA)

- The larvae of these small, grayish-brown moths do all the damage to apple and pear fruit. Adult moths lay eggs on leaves, twigs, or fruit buds.

- Larvae hatch, enter young apples, then tunnel their way to the core where they spend most of their time. When they are ready to pupate, they tunnel out the other side, leaving behind a mess of frass (poop) in the core of the fruit.

- Coddling moths have two generations per year. Each one passes the winter as a pupa in bark or leaf litter.

- They eat the core of pomme fruit, such as apple, pear, and quince.

- To get rid of them, remove leaf litter around the base of the tree and pick up fallen fruit. Cover fruit that is on the tree with nylon footies or paper bags so that larva cannot enter apples.

- Midseason, wrap the trunk of the tree with a band of burlap and Tanglefoot (a sticky substance) or corrugated cardboard to trap larvae as they crawl toward the soil to pupate. Refresh the band in early fall to trap the second generation of larvae as it moves down the trunk.

- Braconid wasps and some trichogramma wasps are natural predators.

- The adult is a tiny (1/4" long) fly that is black with

yellow markings across the abdomen. Eggs are laid singly in punctures in the apple skin—look for little pinpricks. Larvae are white or yellowish.

■ They produce one or two generations per year. The tiny brown pupa (looks like a capsule) overwinters in the soil.

■ Larvae hatch just under the skin of the fruit and tunnel through the flesh, leaving it brown and mushy. Fruit rots quickly.

■ Apple maggots infect apples, blueberries, cherries, and plums.

■ Sticky traps, nylon footies, or paper bags put around fruit act as nontoxic barriers.

APPLE MAGGOTS
(RHAGOLETIS POMONELLA)

- Slugs and snails are hungry, uninvited guests of many city farms. Their whole mission in life is to eat—as omnivores, they love garden vegetables and cat food.

- Most of the slugs and snails on urban farms are nonnative species. They may be troublesome and eat a lot of your garden, but they also provide food for beneficial insects and birds.

- Their bodies consist of one muscular foot. They contract this muscle and scoot along with the help of viscous slime. Two pairs of tentacles retract when they're startled or threatened—eyestalks point up and sensory feelers point toward the ground.

- Slugs have a leathery hood or mantle just behind their eyestalks that protects their internal organs. The hole that opens and closes in the mantle is the slug's breathing hole. Rather than a mantle, snails have a beautiful, whirled shell to protect their vital parts.

- These creatures live under rocks and organic matter. They eat by scraping bits of food into their mouths with a rasping, tonguelike radula.

- Round clear or pearlescent eggs are laid in clusters under rocks, burlap, cardboard, or in leaf mulch. Tiny babies can be seen growing in these jelly balls. They produce many generations each season. Adults hibernate underground.

- Hand-picking is extremely effective; patrol your garden in the evening or early morning. Catch and release them into a wild area where they provide food for ground beetles. Slugs are a favorite food of ducks and chickens as well.

- A copper barrier is effective in keeping slugs and snails out of gardens. Thin cooper strips, 3–4 inches wide, can be wrapped around raised beds, trunks of trees, or containers. When slugs or snails crawl onto copper, they get an electric jolt and won't cross it. Make sure that you don't trap the slugs inside your copper barrier.

LAND
slugs&snails
(STYLOMMATOPHORA)

SPITTLEBUGS

Wat's all that spit on your plants? In May and early June, you may notice globs of white foam on your plants. Don't be alarmed. These are just spittlebugs, the nymph of the froghopper. These cute little green insects don't harm plants, but supply food for garden predators. The foam is a viscous liquid that the nymphs bubble up as camouflage and protection. Spittlebug nymphs do not pupate, but develop wings and become adults as they shed their exoskeleton. Spittlebugs have four to six instars before they are adults. See if you can find the little green bug under all that foam.

Predators and Their Prey

PEST	NATURAL ENEMIES
Aphids	Lady beetles, lacewing larvae, predator bugs, earwigs, parasitic wasps, hoverfly larvae, hornets, yellow jackets, birds
Beetles, destructive	Predator bugs, parasitic flies, nematodes, parasitic wasps
Caterpillars	Predator bugs, birds, earwigs, parasitic flies, ground beetles, lacewings, spiders, parasitic wasps, nematodes
Leafhoppers	Predator bugs, parasitic flies, lacewings, predator mites, spiders, parasitic wasps
Mites	Big-eyed and minute pirate bugs, predator mites, lacewings, midge larvae
Scale	Lady beetles, beetles, lacewings, minute pirate bugs, predator mites, parasitic wasps
Slugs	Parasitic flies, birds, rove beetles, ground beetles

Not all troublemakers have six legs, compound eyes, and an abdomen. Many city farmers must battle larger four-legged pests. No matter what the size, understanding your enemies' habits and habitats will help you know how best to keep them out of your garden.

DEER

- Deer love garden vegetables, flowers, and roses. Fencing that is 10 feet tall or shorter (6') electric fences are effective ways to keep deer out of the garden.

- If you use electric fencing, lure the deer with peanut butter on strips of fabric so that they will get zapped—otherwise they will just leap over fences shorter than 10 feet—electrified or not.

- Cover young tree trunks to protect bark—wrap them with plastic protectors or burlap.

- Plant things that deer don't eat.

- Coyote urine, which can be bought in pellet form at garden centers, is an effective deterrent, but washes away in the rain—great for dry climates.

- Motion-activated sprinklers are extremely effective deterrents.

- Natural enemies include cougar, wolf, and coyote. Watchdogs (or recorded barking) may help keep them away.

RACCOONS

- These critters eat fish from ponds, raid chicken coops, and eat the corn when it is ripe. They will greedily eat all your grapes or other delicious fruit the night before you plan to harvest.

- Cover fruit with tulle netting. Use hot pepper spray on corn. Predator urine and recorded dog barking may be effective. Motion-activated lights have little effect.

- Raccoons are difficult to deter.

- Keep cat food and dog food inside. Lock any pet doors at night— raccoons will come in to eat dog food! Use squirrel baffles to protect bird feeders.

- Raccoon-proof your chicken coop and close it up each night. Provide nighttime protection for other livestock.

- Raccoons are strong and can pull or pry things open to get to food. Urban raccoons are not scared of humans and are often fed because they are "cute." Don't feed raccoons!

- They typically have one litter of four to five babies each spring.

RABBITS

- Feral rabbits are a problem for many city farmers. They eat everything in the garden. Nearly 80 percent of their diet is grass, but you would never know it after they have thoroughly consumed your lettuce and broccoli.

- Rabbits have three to four litters of six babies each year.

- They feed in the early morning, but can be seen feeding throughout the day (when populations are large).

- Grow greens in a covered hoop house.

- Silver mylar tape or old CDs hung on strings around beds are an effective deterrent, since rabbits frighten easily. Fences should be at least 3 feet tall with wire buried underground to prevent rabbits from digging through to the garden.

- Catch rabbits in humane traps using carrots, lettuce, or broccoli as bait. Take them to a shelter to be neutered—there are rescue agencies in many big cities that will provide sanctuary for feral rabbits.

- Their natural enemies include hawks, falcons and other large raptors, feral cats, coyotes, and dogs.

SQUIRRELS

- Mischievous thugs of the city farm, squirrels dig up and eat fall bulbs, sunflowers, squash, and pea and bean seeds. They vandalize tree fruit and berries. Nuts are buried all around the garden, but forgotten and then sprout in the spring. Peanuts, misplaced by the local squirrel, sometimes sprout mysteriously in my garden.

- They have one to two litters of three to five babies each year.

- They are crafty, persistent, and hard to stop. Use squirrel baffles on bird feeders. Cover fruit or nuts with tulle netting.

- Try hot-oil spray on sunflowers or a quick spray from the hose, when you are outside, to chase them away.

- Cover new seedbeds or flower bulbs with chicken wire or a floating row cover.

- Squirrels have no natural enemies—they're too quick for most cats or dogs.

RATS
AND OTHER RODENTS

- Urban rodents find food, water, and shelter on city farms. They like to eat seedlings, seeds, and fruit. They love to make a dry, fluffy compost bin their home.

- These rodents produce three to four litters of four to seven pups each year.

- Some burrow and live in rock gardens, woodpiles, or basements, while others live in trees and attic crawl spaces.

- They are omnivores and will eat just about anything—they like to eat the leftovers in the chicken coop and even chicken poop!

- They can fit through a hole ½ inch square or larger. Cover potential entry holes with ¼-inch hardware cloth or stuff them with steel wool.

- Keep food harvested. Keep compost wet and turned frequently to disrupt rat habitat. Feed chickens only the amount they can eat.

- Cats, terriers, and urban raptors are their natural enemies.

CROWS
AND STARLINGS

- They eat plenty of bugs in the garden.

- They love to eat seeds and tender seedlings.

- Protect your seeds or young seedlings with a hoop house or floating row cover. Mylar flash tape, strung around a bed, may startle them and keep them away.

- Scarecrows (meant to fool the birds into thinking there is a person out in the garden) have been used for centuries to deter these birds. I'm not sure they fool the urban crow, however.

BEARS

- These are not a problem for most city farmers, but we have had sightings within the Seattle city limits. Many outlying communities are experiencing sightings each spring as bears come out of the hills looking for food in garbage cans.

- Bears dig up the compost bin or destroy the worm bin to get at food. They will eat blueberries, blackberries, and root crops in your garden.

- Be careful; don't go out when you see a bear.

- Keep compost bins and garbage cans inside. Recycle yard waste through curbside pickup if bears dig up your compost bin. Don't compost food waste.

- A motion-activated dog-bark recording or an electric fence may serve as an effective deterrent.

KEEPING CATS AT BAY

Cats love nothing better than a freshly seeded garden bed to use as a litter box. Keep them out of your new beds by laying chicken wire over the soil or crisscrossing small sticks across the area. Frustrated city farmers stick bamboo skewers like a pointy jungle trap to keep squirrels and cats from digging in freshly dug beds. Moist soil is less appealing, so keep the surface of the soil moist or drape a layer of floating row cover over the bed until your seeds have grown some.

MOLES AND GOPHERS

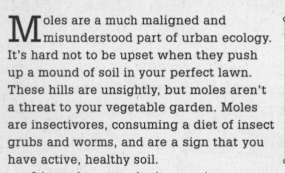

Moles are a much maligned and misunderstood part of urban ecology. It's hard not to be upset when they push up a mound of soil in your perfect lawn. These hills are unsightly, but moles aren't a threat to your vegetable garden. Moles are insectivores, consuming a diet of insect grubs and worms, and are a sign that you have active, healthy soil.

It's gophers you don't want in your garden. Gophers eat plants, especially root crops. You can tell by the shape of the mound if you are dealing with a mole or a gopher. Mole mounds look like volcanoes with the top of the hill centered. Gopher mounds are fan shaped with the top off-center.

Trapping either critter is difficult—remember Bill Murray from Caddyshack? Dogs and cats that are good hunters can keep these populations in check. Gophers love vegetable crops—bury ½-inch hardware cloth 18 inches deep and 8 inches high to keep gophers out of the garden. Predator urine, which can be bought in pellet form at garden centers, might help keep gophers out of the garden as well. Crushing mole hills and tunnels as soon as you see them can push their activity to the periphery of your yard, where it is more acceptable.

Managing weeds is no different than handling
pests or larger creatures. Learn the plant and its ways.
Find out if it is useful or beneficial. Use the best, nontoxic
technique for removing or controlling it.

Bindweed, field
(CONVOLVULUS ARVENSIS)
or common
(CONVOLVULUS SEPIUM)

- Often called morning glory, this perennial vining plant spreads by roots and likes acidic, neglected soil. Plants produce vines up to 30 feet long with heart-shaped leaves and large white (common) or small pale pink (field) trumpet-shaped flowers. Thick, spaghetti-like roots spread near the soil surface, are brittle, and break off easily—each piece of root will grow a new plant.

- Their brittle roots make it tricky to eradicate this weed. If your soil is too hard to loosen and remove roots without breaking them, sheet-mulch the area to make digging easier. The root system under the mulch will be massive but easy to pull.

- Keep digging and removing any piece of bindweed. Sifting the soil to remove absolutely every last piece of root is extremely effective.

- If its roots are embedded in bricks or concrete, or entwined in tree roots, sheet-mulch to keep growth under control and pull the new shoots as they emerge.

- Do not add these weeds to compost bins. Add lime to make the soil less acidic.

- Purportedly, the roots of common bindweed can be chopped and boiled for a bland, but nutritious meal (I have not tried this). Vines make great ties for herb bouquets, cordage for baskets, and are fantastic crowns for scarecrows or children!

CHICKWEED

(STELLARIA MEDIA)

- This lovely early spring and fall annual weed with small lobed leaves and white star flowers is one of the most common weeds in garden beds or planting areas.

- Chickweed is a great indicator of soil health. If it is green and robust, you have fertile soil ready to grow vegetables. If it is anemic and woody, your soil lacks nutrients and needs some compost. Chickweed dislikes hot, dry, and compacted soil.

- This weed spreads by seeds that form on flowers almost as fast as the flowers appear.

- Pull the whole plant as flowers appear and before the seeds set. It's a great source of nitrogen for your compost bin or greens for the chickens.

- Tender, young leaves, stems, and flowers can be cut with scissors and added to salads or sandwiches. They have a fresh green-grass taste. Older plants are too dry and woody to eat.

- This perennial, relentless, spreading weed thrives in wet, poorly drained clay or silty soils.

- It spreads by runners, like strawberries or spider plants, that form roots and make a new baby plant.

- Creeping buttercup is easy to dig up. Loosen the soil with your digging fork and pull out the plant—roots and all. Sheet-mulching is an effective way to slow growth, though buttercup may grow through thin layers.

- Do not add it to your compost bin, but it's OK to throw to the chickens.

- It's not edible and is mildly poisonous (though not deadly). Except for the "Do you like butter game," there is no use for it in your garden. If you have children, don't tolerate buttercup near edibles, since it looks like parsley and eating even a small bit can result in a wicked stomachache.

Creeping BUTTERCUP

(RANUNCULUS REPENS)

DANDELION

(TARAXACUM OFFICINALE)

- This perennial herb with a long taproot is very adaptable and will grow in heavy clay and compacted soil. It can tolerate acidic soil and is the most common weed in the lawn and garden.

- Loosen soil and pull roots before flowers have made seeds. Use a pronged dandelion puller tool to pop plants out of the ground when soil is moist.

- Keep roots and seed heads out of the compost.

- All parts are edible—cooked or raw. Leaves can be added to salads or steamed—they will be less bitter if harvested young and given a steady supply of water. Flowers can be steeped to make tea or wine; roots can be peeled, sliced thin, and cooked with other roots. It is high in vitamin A and calcium.

DOCK

(RUMEX)

- This perennial member of the buckwheat family grows in sun or shade in poorly drained, waterlogged, or acidic soils.

- It's commonly found in lawns, garden beds, and neglected corners. It spreads by seed.

- Cut flower stalks before they set seed. Dig out roots when the soil is soft and wet. Sheet-mulching is effective for killing dock.

- New center leaves can be picked and eaten raw or cooked. They are slimy and delicate with a fresh, slightly tangy taste. Young roots can also be eaten.

- Brown seed stalks are a striking addition to cut-flower arrangements.

- This perennial has been around since the age of the dinosaurs. No other weed is as hated by gardeners. It is extremely adaptable and thrives in moist soils—common in boggy areas and in heavy clay soils.

- Horsetail spreads by a tuberous root system.
- Pulling just stimulates growth and makes more plants. Roots are fine, the color of soil, and hard to dig out. Cut stems at the soil line to slow their spread. Sheet-mulching will not stop horsetail, which can grow up through asphalt.

- Don't put this weed in the compost pile or feed it to livestock.
- Not edible. Greens can be used as a crude scouring pad or steeped to make an antifungal soil tea.

HORSETAIL

(EQUISETUM ARVENSE)

Lambsquarters

(CHENOPODIUM ALBUM)

■ This annual edible weed is a member of the spinach family and grows in cultivated soil. When grown in fertile soil it will be enormous (over 6 feet tall and 4 feet wide). Plants that are small and lack vigor indicate that soil has low fertility.

■ Lambsquarters spreads by seeds.

■ Pull the whole plant (before it sets seed) and add it to compost or feed it to livestock.

■ Magentaspreen lambsquarters is an attractive edible that can be intentionally sown with bedding plants.

■ A delicious and highly nutritious green, it is a welcome weed. Leaves are a rich, buttery addition to salads or sandwiches. Lightly sauté leaves and flower buds in olive oil for a satisfying side dish.

Pigweed
(AMARANTHUS RETROFLEXUS)

- An annual weed that is related to amaranth, pigweed is a common weed of the mid- to late-summer garden. It thrives in healthy soil or grows less vigorously (but still persistently) in soil that lacks nutrients.

- Pigweed spreads by seeds.

- It's easy to pull whole plant; get to it as soon as green flower buds start to appear.

- Add it to the compost pile before it sets seeds. Feed it to livestock after seeds start to form.

- Leaves can be eaten raw or cooked, but are bland and forgettable.

- A tenacious perennial medicinal herb, plantain appears as a narrow-leafed version (English, plantago lanceolata) or as a broadleaf (plantago major); both have similar habits and habitats.

- Plaintain grows in heavy clay and poorly drained soil. It is common on the edge of footpaths and in lawns and cultivated soil.

- An extremely hardy weed, it can withstand foot or car traffic, and spreads by seeds that form on Qtip®-like flower stalks that are covered with white flowers.

- An unimposing weed, it can share space in any of your garden beds. Pull the whole plant and add it to your compost pile or feed it to livestock.

- Plaintain is an important plant for healing the bug bites or stings of summer. Chew leaves into a pulp and spit them out onto a sting or itchy bite. Leave the pulp on the skin for 10 to 15 minutes. Wash the area with cool water and mild soap—the redness and itch should be gone.

- Dried seed husks are used as a homegrown fiber laxative.

PLANTAIN

(PLANTAGO SPP.)

THE Prickly AND THE Soft: KIDS LOVE WEEDS

As you use your senses to explore the garden, be sure to touch the soft and the prickly. Many weeds that vex adult gardeners have a special appeal for children. The soft leaves and the giant flower stalks of mullein which produce gazillions of seeds are a wonder to small people. The soft leaves are great carpets or beds for bug and fairy houses; the dried flower stalks make cool torches when dipped in wax (do this with an adult's help).

Dangerous and prickly plants are always fascinating for kids. Green, sticky burdock burrs (which were the inspiration for Velcro fasteners) are great for holding small flower boutonnieres or for makeshift dart games.

Teasel, a common field weed, is a water fountain plant when green and the "most dangerous plant" when it dries out. The green leaves of teasel attach to the stem, creating a cup that holds water and creates little pools from which insects and birds can drink. As teasel dries, it sports sturdy thorns along the stem and pincushion spikes on the flower head. Let a few weeds grow—unexpected surprises may be in store!

Common Diseases

Plant diseases have habits and habitats, too. Learn about diseases so you can implement the most effective controls.

Powdery mildew

- This light-colored, powdery spore grows on shoots and both sides of leaves.

- Powdery mildew affects many different plants and fruit trees. There are several different powdery mildew fungus species.

- Fungi spores are spread by wind and grow on the surface of leaves. Fungi love temperate humid environments. Plants that are not in the right place, suffer from neglect, or grow in depleted soil are highly susceptible.

- This disease is brought on by moderate temperatures and shady conditions. Stressed plants (those that have not gotten enough water) are also at risk.

- Almost every vegetable and fruit in your garden could get powdery mildew, which affects leaves but not fruit or vegetables.

- New growth on grapes and fruit trees is attacked—new growth is dwarfed or misshapen, and new fruit can develop russetted scars or cracks.

- Grow mildew-resistant varieties. Prune plants to increase air circulation. Encourage an active insect and bird community. There are bugs that feed on this fungus.

- Plant in a sunny location. Use soaker hoses or drip irrigation to keep moisture off leaves. Build healthy soil and topdress heavy feeders, such as squash and cucumbers, with compost every three or four weeks. Remove infected branches and dispose of them in the trash or as curbside yard waste.

Sterilize tools to keep the disease from spreading.

- Try a baking soda spray to control this disease: Mix 1 quart water, 1 tsp soda, a few drops of dish soap, 1/2 tsp vegetable oil. Spray this on leaves at the first sign of disease.

Fungal blight

- Often called late blight, this is a fungal disease that affects tomatoes and potatoes. Late blight was the main culprit in the disastrous failure of potato crops in Ireland in the 1840s.

- The blossom, or bottom, end of tomatoes develops brown spots and fruit rots. Blight may appear as brown, sunken areas on the blossom end of fruit, or dark tan or brown spots on leaves and branches.

- Fungal blight thrives in high humidity or moisture and temperatures around 68 degrees, and it's common when there is a cool, wet period in September.

- This disease affects leaves and fruit. Three weeks after infection, the whole plant can be defoliated.

- Use drip irrigation or soaker hoses to keep water off leaves. Pinch off the bottom leaves. Prune to increase air circulation so the plant dries out more quickly. Grow tomatoes in high hoop houses (6–8' tall).

- Sanitation is essential for stopping the spread of late fungal blight—fungi overwinter in infected fruit and plants. If disease is present, clear and dispose of infected fruit and plants. Disinfect tools.

Blossom end rot

- In this disease, the blossom end of fruit turns yellow, then brown, and starts to rot.

- It's caused by a calcium deficiency and inconsistent watering (periods of drought followed by lots of water) while fruit is developing.

SANITATION IS ESSENTIAL

- Among the plants affected are tomatoes, peppers, squash, cucumbers, and eggplant.

- Water consistently, and use a fertilizer that is balanced or is low in nitrogen. Build healthy soil; maintain the proper soil pH for optimal nutrient uptake. Add a sprinkle of gypsum; this will help make calcium more available.

- Remove affected fruit. Topdress with high-quality compost.

Brown rot

- A powdery brown rot that affects stone fruit, such as plums, peaches, and nectarines. It attacks fruit blossoms, causing them to wither, and can kill young twigs. As fruit develops, the fungi envelop ripening fruit, leaving a moldy, withered mummy.

- Sanitation is essential: Remove and dispose of any affected fruit or branches. Fungi overwinter on mummies, affected twigs, or blossoms.

- Keep irrigation off blossoms, foliage, and fruit. Plant disease-resistant varieties. Prune and thin fruit to increase air circulation. Build soil and fertilize adequately.

Distinguishing Friend from Foe

Getting to know the creatures on your city farm is amazing. Learning about the habits and habitats of creatures saves a lot of work and worry. Letting nature keep the balance and keenly observing nature in action constitute a great adventure.

SMALL PLOT FARMING

A remarkable amount of food can be grown on a surprisingly small plot of land—just ask the folks at SPIN Farming, a company that teaches others how do to just that. SPIN, or Small Plot INtensive Farming, was designed to teach those with small holdings, such as city and suburban lots, to grow enough food to sell at local markets.

The company sells guides that walk you step-by-step through the process of changing your lawn into a valuable source of homegrown food, even extending your season at the beginning (earlier planting) and end (later harvesting). A calculator on SPIN's Web site helps you figure out how much space you have and just how much money you could make by growing and selling your produce locally.

Philadelphia farmers Steve and Nicole Shelly have put the SPIN theory into practice. Their 280 beds turn an average of three to four crops

each season, and they grow 100 varieties of 50 different types of vegetables. They make more than $50,000 within a nine-month growing season.

"What we're trying to do is recast farming as a viable small business in the city," says one of the founders, Roxanne Christensen. "SPIN applies small business practices to farming, and its subacre growing methods make it possible for many more people to follow their desire to farm."

LEARN MORE:
spinfarming.com OR
spingardening.com

EXTENDING THE HARVEST

Get the most out of your garden by extending the length of time that you eat from your city farm.

WO THINGS YOU CAN DO TO make it last: (1) Grow a winter garden using heat-trapping devices and cold hardy vegetable varieties or (2) preserve your harvest and eat from your pantry all winter long.

Fall and Winter Gardening

Winter gardening isn't for everyone. Many gardeners are relieved when the growing season ends. Although they love and value growing their own food, it is a lot of work to keep things alive—to water, harvest, and eat everything they have planted. Even if year-round gardening is not in your farm plan, you can use season-extension techniques at the beginning and end of your growing period to eke out more food from your city farm. For tips specific to your area, check with your local cooperative extension service.

There are two ways to garden during fall and winter. You can grow crops that will be eaten in fall and early winter or you can grow overwintering crops that you will harvest in early spring. Fall and winter crops are sown in July and August directly into garden beds. They will grow to maturity and be ready to harvest beginning in late September. Crops that overwinter are sown in summer to be transplanted as starts in September—these tiny plants will sleep all winter and start growing in spring for an early harvest.

Making your winter garden

Locate your winter garden in the warmest, sunniest microclimate you can find. An ideal spot faces south or west and is gently sloped to capture heat. Areas against a wall or fence may absorb more heat and protect plants from winds. Raised beds warm up faster than those that are sunken or at ground level. Dark-colored containers, protected from wind, gather more heat than lighter-colored materials.

Prep winter garden beds with an inch or two of compost or worm castings—don't add fertilizer, since you don't want to stimulate too much growth. In the spring, water overwintering vegetables with liquid fertilizer as new growth begins to appear and repeat every four weeks. Give seeds and seedlings just a little more space than is usually recommended (see chapter 7) since, in colder soil, there is less microbial activity, making it harder for plants to find nutrients. Giving plants a bit more space will eliminate competition for nutrients. Warm the soil and protect winter crops by growing them under cover.

Hardy varieties

Many vegetables are cool-season crops, including lettuces, endive, radicchio, mache, members of the cabbage tribe, and legumes. These plants are typically hardy and will grow through the winter with a little protection. Look for cold-hardy varieties or heirlooms that come from cold climates like Russia. Ask around at the farmers' market. Local farmers grow hardy varieties to satisfy the tastes of city dwellers.

An easy green to try first is mache, or corn salad. This winter salad green comes from Europe and grows wild in the stubble of grain fields (hence the name corn salad). It is a delicious and buttery green that is popular in Europe but is seldom seen here. Mache grows in small heads that are too laborious to harvest commercially—it is a delicacy that city farmers who garden in fall and winter can enjoy. Sow seeds in August for winter greens or grow seedlings indoors and plant out in spring as soon as the soil can be worked.

Arugula will grow during the cold season with minimal protection. Brune d'hiver, winter density, marvel of four seasons, red deer tongue, and Australian yellow leaf are reliable and attractive hardy lettuce varieties for fall and winter harvest. Or sow some of your leftover lettuce or brassica seeds in rows under a hoop house and see what happens.

Timing

Gardening year-round is tricky and can be challenging for even the most committed city farmer. The key is timing. Fall and winter gardens are planted in the summer and harvested when it gets cold.

Crops to be harvested in fall and winter are sown as seeds directly into garden beds in July and August. Overwintering crops are sown in pots in August and transplanted in September. Many members of the cabbage tribe will overwinter as tiny plants. Overwintering plants should be very small—just two pairs of true leaves—so that less plant tissue is exposed to cold and frost. These plants will stop growing during the winter and will start growing again as soon as the soil starts to warm up. Cilantro can be sow in late August to overwinter and the new growth in spring will be lush and slow to bolt. You may get three or four harvests from this early crop.

Getting seeds to germinate in midsummer weather is tricky. Soil dries out rapidly and little tender sprouts can wither and die quickly. Keep soil and seeds evenly moist by draping a layer of burlap or row cover cloth on the soil and watering it. Lift the burlap to make sure the soil is getting evenly moistened. Remove the burlap as seeds germinate and keep the bed wet as sprouts grow. Or find a partially shaded spot to raise seedlings in pots (to transplant in September).

First frost date

This marks the end of the summer growing season. Knowing your first frost date (or the start of cold, wet fall weather) will help you know when you need to get the last of the summer fruit off the vine. In Seattle we like to leave the tomatoes on the vine as long as we possibly can, so more will be vine-ripened. As a result, we keep a close eye on that date. The quality of summer fruits deteriorates with cold, wet weather, and fungal diseases can set in.

Planting for fall and winter harvest requires an appreciation of the first frost date for your area. Crops that are planted in July and August can be harvested in late fall. Tiny broccoli or cabbage seedlings that are transplanted in September will overwinter for a burst of growth and an early harvest in the spring.

Microclimates affect which areas see frost first and which areas warm up first. Look for frost pockets and places where snow stays longest to identify your coldest microclimates. Many vegetables can tolerate mild to hard frosts or even a blanket of snow.

To find the first frost date in your area, go to the Web site of the National Climatic Data Center (cdo.ncdc.noaa.gov) and search "freeze/frost data" for your region. Your local cooperative extension service will have frost date information, or type "first frost date [YOUR TOWN]" into your favorite search engine.

Growing under cover

Simple season extenders can be used by city farmers to protect plants and warm up soil. Covering plants traps heat; protects them from damaging rain, wind, and frost; and keeps the garden growing longer.

Planting Calendar

CROPS TO PLANT IN JULY AND AUGUST FOR FALL HARVEST	CROPS TO PLANT IN LATE JULY AND AUGUST FOR FALL HARVEST	CROPS TO TRANSPLANT IN SEPTEMBER FOR SPRING HARVEST (PLANT SEEDS IN JULY/AUGUST)	CROPS TO PLANT IN OCTOBER FOR SUMMER HARVEST
beets	beets	beet greens	cereal grains
heading broccoli	cabbage	sprouting broccoli	fava beans
broccoli raab	carrots	cabbage	garlic
cabbage	cilantro	collard greens	overwintering onions
carrots	collard greens	endive	shallots
cauliflower	endive	kale	
cilantro	kale	lettuce	
collard greens	lettuce	mustard greens	
endive	mache	overwintering onions	
kale	mesclun mixes	radicchio	
lettuce	mustard greens	Swiss chard	
mesclun mixes	parsnips		
mustard greens	spinach		
parsnips	Swiss chard		
radishes			
snow peas			
spinach			
Swiss chard			

Cloches

A cloche is a simple covering that captures the maximum sunlight from all angles. The temperature of the air and soil can be raised by 5–10 degrees Fahrenheit when covered by a simple cloche.

A cloche is a classic season extender first used by Parisian market gardeners in the late 1800s. It is a hive-shaped glass or plastic bell that is placed over individual plants to warm soil, intensify the effect of the weak winter sun, and protect plants from frost. You can easily make your own inexpensive cloche out of anything that transmits light. For example:

- *Plastic milk jugs with the bottoms cut out*
- *Recycled windows leaned together, like a teepee*
- *Plastic draped over branches and anchored at the edges*
- *Glass domes*

Hoop houses

A hoop house is a mini-greenhouse constructed out of plastic or wire hoops with plastic or a floating row cover draped over and pulled tight. These tunnels are easily constructed.

Make a hoop house for a 3 x 8 foot bed using plastic pipe as follows:

MATERIALS

3 hoops, made from 7- to 8-foot lengths of PE (Polyethylene) pipe

10 x 14 foot heavy (6 mil) greenhouse plastic— or similar clear, heavy plastic

6 rebar stakes in 2-foot lengths

9 cloche clips—for holding plastic to PE pipe

8 large rocks or bricks— for weighting down the plastic

1. Pound rebar into the ground at the corners and middle sides of your garden bed. Leave them 10 to 12 inches above ground.

2. Slide PE pipe over the rebar.

3. Drape the plastic over the hoops, pulling it taut.

4. Secure the plastic with three cloche clips on each hoop and anchor the edges with rocks or bricks. For added stability, tie a long bamboo pole or stake along the tops of the hoops.

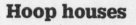

Make a hoop house out of heavy-gauge wire for a lightweight plastic or floating row cover as follows:

MATERIALS

4–5 hoops, made from 6-foot pieces of heavy-gauge wire (like coat hanger)

10 x 14 foot lightweight clear plastic or floating row cover

12 or more clothespins, for affixing plastic or row cover to hoops

8 large rocks or bricks, for weighting down the plastic or row cover

1. Push one end of the wire 12 to 18 inches into the ground at either corner (and across the middle) of your garden bed.

2. Space hoops evenly across the bed.

3. Drape the plastic or row cover over the hoops; pull it taut.

4. Secure it with 4 clothespins per hoop and pin down the edges with bricks or rocks.

Cold frames

Cold frames offer the benefits of both a cloche and a greenhouse in terms of protection from chilly temperatures. A cold frame is a box with solid sides that is open on the bottom and fitted with a sloped glass top. Cold frames are very sturdy, but slightly less mobile than a cloche due to their size and weight. Using recycled materials and a few power tools, you can easily construct a cold frame.

Cold frames can be used in a number of ways to protect crops and extend your growing season. Use a cold frame to harden off spring seedlings that have been started indoors. Put entire flats of plants in a cold frame to grow them under natural light and acclimatize them to outdoor temperatures. Regulate temperatures in your cold frame by opening the lid.

Sow spring greens or root crops in your cold frame. Warm the soil a week before planting seeds by placing the cold frame over the bed where you will plant. Sow seed directly in the soil inside the cold frame and close the lid. The seeds will germinate more quickly in this warm environment. Remove the cold frame when the seedlings are more mature and the temperatures are warmer. Cover transplanted starts as you would with a hoop house or cloche. Since cold frames are usually only 12 or 18 inches tall, they only work while transplants are small.

Greenhouses

Greenhouses are the ultimate season-extension tool. For those with space and enough money, you can grow food year-round in a heated greenhouse. Most greenhouses are houselike structures covered with plastic, glass, or special glazing that allows light to penetrate from ceiling to floor. With a greenhouse, your seedlings can grow with natural light. Some greenhouses have open floors so that certain crops, like tomatoes, peppers, or eggplant, can grow the entire season inside. This is an effective technique for cooler climates where it can be challenging to grow heat-loving crops. Unheated greenhouses also provide plenty of protection and can be used to grow seedlings that will be moved to the garden when temperatures are right.

Other ways to keep plants warm

There are other ways to warm up the soil and protect plants from cold as well.

- *Wall o' Water® is a popular season extender. It is a round collar with chambers, filled with water, that is put around transplants to create a warmer growing environment. You can make your own by filling liter bottles with water and placing them around your plant.*

- *Red and black plastic film can be laid over garden beds to warm the soil before sowing seeds or transplanting.*

- *A floating row cover can increase soil and air temperatures by 5 degrees Fahrenheit and protect crops from pest infestations.*

- *Raised beds constructed with walls of rock, brick, recycled concrete, or other heat-retaining materials will slowly release heat accumulated during the day to warm the soil even more.*

- *Since dark materials absorb more heat, adding dark compost to your soil or burlap sacks covered by dark plastic can raise soil temperatures considerably.*

When growing under cloches and cold frames, pay special attention to temperature, humidity, and soil moisture. Except during long freezing periods, open (or vent) your cold frame or cloche during the day and close it at night. Venting helps control temperature and humidity. Getting air to your plants also helps to prevent fungal outbreaks.

If you are using a hoop house, lift the plastic on either end so that air has a chance to pass through. On cloudy or cold days you may not need to provide much ventilation. Open both ends of the tunnel on sunny winter days so little seedlings don't fry.

Prop the lid of your cold frame open to adjust ventilation and temperature. Start by opening the lid a few inches on overcast days, and 2 feet on sunny days; open it all the way or remove it when the weather becomes warm. Since hot air rises, some plastic bell cloches have small vents in the top that can be adjusted.

Plants grown under cover will get water only if you provide it, so don't forget to check soil moisture (chapter 8). When a cloche or cold frame is closed, moisture that would escape into the air condenses on the glazing and returns to the soil. During damp and cloudy periods, you may not have to water as much. When you vent your cloche or cold frame, you lose moisture to evaporation and you will need to water more.

Getting the hang of planting a winter garden takes practice. If it seems overwhelming to plan an entirely new garden for winter, start with cover crops (chapter 3). As vegetables stop producing, sow grain or legume cover crops. This can be your initial winter garden. Next season, you will be eating the benefits of this soil building.

Making It Last

Eating from your urban farm year-round

Preserving food is another great way to make your harvest last. If your winters are icy and gardening isn't an option, eating out of your pantry or freezer is an ideal way to extend your harvest.

Preserving produce

There are three methods of preserving garden produce commonly used by city farmers—dehydrating (also known as drying), freezing, and canning. Not every method works well for every vegetable or fruit. Some things are better when they are dried, while others are best when canned. Culinary herbs and fruit can be dried for later use and bumper crops of vegetables can be blanched and frozen to add to winter soups. Canned jams and syrups can be given as gifts or opened in midwinter as a sweet reminder of summer. You will find the ways to preserve your harvest that work best for you, your taste buds, and your cooking habits. Use only really fresh, top choice, ripe produce at the peak of freshness. Taste or texture will not improve with processing.

DRYING

Drying food is fun and easy. Humans have been drying food to preserve it for eons. Put simply, drying means that all the water is removed from vegetables or fruit. Without water they won't rot, so they can be stored for a long time. Dehydrated foods are very efficient, since they keep for a long time without any added energy.

It is possible to dry things by simply spreading them out on newspaper or a cardboard box, in a paper bag or on window screens on the balcony. You can also dehydrate produce on racks or trays in your oven, set to its lowest setting.

However, nothing beats an actual food dehydrator. Food dehydrators have a heating element at the base. With or without a fan, warm air rises through screened trays where food is placed. Some models have round, plastic towers with doughnut stacking trays; others have square boxes with framed screens that stack above the heat source. Discarded food dehydrators can sometimes be found at thrift stores, garage sales, or at online auction sites.

Label Everything!

Inevitably I find several mystery containers in the back of the freezer at the end of every season. Since I don't know how old they are, these end up going in the worm bin or to the chickens—what a waste! Or I dump in what I thought was Swiss chard, only to find it is celery, completely changing the taste and texture of the dish I am preparing.

You aren't finished preserving your harvest until stuff is labeled. Keep a roll of masking tape and a Sharpie® marker handy and label every container

Drying fruits and vegetables

Preserving your farm produce by dehydrating it is a snap. Here are some basic guidelines:

- *Blanch—most vegetables to stop enzymatic action.*

- *Keep pieces thin—$1/8$–$3/8$ inches to dry more quickly.*

- *Check daily—food should be dry, but not crispy fried.*

- *Dip fruit—lemon water will keep food from turning brown.*

- *Switch trays—bottom trays near heat may dry too quickly, while the top tray dries more slowly.*

- *Don't cook—take the water out by keeping the temperature between 95 and 130 degrees F.; use a thermometer to regulate the heat.*

- *Rehydrate—dried fruit and leathers can be eaten as they are. Rehydrate other vegetables by soaking them in boiling water or simmering them on the stove for 20–30 minutes.*

- *Store dry foods in a cool place at 50–60 degrees—they will last a year or longer.*

you put by—it doesn't matter if it has been canned, dried, or frozen. Specify what's inside and the date it was prepared. Then there will be no surprises and less waste. You can also write down an expiration date on your containers. Then you will know what needs to be eaten sooner or what will last longer.

FREEZING

This is my favorite way to put food by. When I started eating locally and preparing all my own meals years ago, I bought a used upright freezer from an appliance recycler. Initially, I just froze jam and fruit, but I soon learned that a lot of my garden produce could be easily frozen. I began cutting up vegetables so they were ready to go into anything I was cooking for dinner.

Not everything that comes out of your garden is suitable for freezing. This isn't an option for vegetables that are eaten raw. But it is a great way to preserve fruit and any vegetables that will be cooked before eating. Hearty greens such as kale, collard greens, and Swiss chard actually become more nutritious after they

[Recipe]
Blanching and Freezing Vegetables

1. Wash and cut vegetables into the desired shape.

2. Put prepared vegetables in a strainer and submerge the strainer in boiling water. Blanch for 2 to 4 minutes, depending on the vegetable.

3. Submerge in an ice-water bath to stop them from cooking any further.

4. Drain. Remove excess water by rolling the vegetables in a towel to dry.

5. Pack them in plastic freezer containers or put them into double freezer bags. Remove air, but allow a little space for food to expand when it freezes.

6. Label the containers or bags.

have been blanched and frozen before being cooked in a meal.

The downside of freezing is that is takes constant energy to keep food preserved and the texture of food is changed when it is frozen and then thawed. Experiment with freezing vegetables by themselves or in prepared dishes.

Straight to the freezer

Fruit can go directly into the freezer either whole or sliced. Onions, leeks, shallots, and garlic can be chopped, diced, or minced; put in freezer bags; and placed straight in the freezer. Having alliums prepped and ready to go makes meal preparation fast and easy.

Slip the Skins Off!

Quickly remove the skins of tomatoes and stone fruit—plums, peaches, nectarines, and apricots—before you process them. Submerge them in boiling water for 30 seconds to 1 minute or until skin blisters, then cool in a ice-water bath. Skins should come off easily.

ENZYMES AT WORK

If you freeze most vegetables without first blanching them, they will thaw into an unrecognizable, horrible-tasting, cardboardlike substance. That's because enzymes in vegetables break down vitamin C and change sugar to starch. These are dormant when the produce is cold, become more active when the vegetable is warmed up, and die in extreme heat. Submerging vegetables in boiling water for 2 to 4 minutes kills the enzymes, making the vegetables ideal for freezing. Blanched frozen vegetables will taste great for up to 9 months.

CANNING

Preserving food by canning is a farmstead classic. Special equipment is needed to can most vegetables and meat. Fruit and high-acid foods can be canned at home without too many specialty tools.

Be careful and diligent when you are canning—botulism can form during the canning process, and it kills. Keep everything spotlessly clean and boiling hot. When canning, cleanliness is next to godliness. Follow your recipe precisely. Start with clean jars, new lids and rings, and high-quality produce.

Hot water bath-canning

If you have never canned anything before, the process may seem mysterious and scientific. There is some science to it; you want to make sure that conditions are not right for botulism to grow. To be safe, follow your recipe precisely, process only fruit and high-acid foods, and sterilize everything in boiling water before you use it.

Your recipe will give specific instructions. Here is the basic procedure for "hot pack" boiling water bath-canning:

1. Put all mason jars, lids, and rings in near-boiling water until you need them; heat water in a saucepan or teakettle so you will have extra if needed. Keep everything super clean.

2. Fill the canner with water and bring it to a boil.

3. Put hot liquid and food into the hot mason jars.

4. Use a wooden chopstick or rubber spatula to remove air bubbles, leave 1/4 – 1/2-inch space at the top of each jar (called headspace).

5. Wipe the rim with a clean hot cloth. Put hot lids and rings on—tighten rings.

6. Arrange jars on the canning rack and lower the rack into the canner—there should be 2 inches of water covering the jars; add more boiling water if needed.

7. Bring back to a boil and process for the designated time.

8. Remove the hot jars with a can lifter and cool them on a clean towel.

9. Listen for the lovely "ping!" as the lids seal. If lids are depressed, you are good to go. If they don't seal, refrigerate those jars and eat right away.

10. Label everything and display them on a shelf for all to admire!

HOT WATER BATH TOOLS

You will need a few pieces of equipment in order to can your farm produce in a hot water bath. You may have some things in your kitchen already; the rest can often be picked up at garage sales, thrift shops, or at online auction sites.

- Large canner with jar rack
- Large stockpot
- 1 or 2 saucepans
- Mason jars
- New lids and rings
- Canning funnel
- Can lifter
- Tongs
- Clean towels and washcloths
- Labels and Sharpie® marker
- Oven mitts and hot pads

Storing Your Harvest to Eat Fresh

After you have harvested your bounty, store fruits and vegetables properly so that everything stays at the peak of freshness and flavor.

Here are some basic storage tips:

Keep your refrigerator between 34 and 40 degrees F. to stop bacterial growth and to slow ripening and spoilage.

- *Don't wash vegetables or fruit before putting them in the refrigerator. Excess moisture quickly deteriorates the surface of fruits or vegetables causing them to rot more quickly. Wash garden produce before you prepare or eat it.*

- *Bag foods that lose moisture quickly—use plastic or paper.*

  - *Remove tops of beets, carrots, parsnips, and radishes before storing. Root vegetables will keep for several weeks if the tops are removed. Edible tops should be eaten in 3 – 5 days.*

    - *Let some fruits ripen on the counter before you put them in the fridge.*

    - *Tomatoes or anything picked to ripen after harvest shouldn't be refrigerated.*

  - *Revive wilted greens by submerging them in a bowl of cold water and putting them into the refrigerator for a few hours.*

VEGETABLE AND FRUIT LIST

Keep your harvest fresh and preserve it using the method that works best for you. Here are recommendations about how to store produce for eating fresh, as well as how to dry, freeze, and can. Also included are some dandy recipes from family and fellow city farmers.

Vegetables

Beans—bush, pole, runner, snap, wax/dry

FRESH. Store in a plastic bag; keep refrigerated; eat within 5 days.

DRY. Wash and cut into desired size; blanch for 4 minutes; dry until brittle; will keep 8 months.

FREEZE. Cut into desired size; blanch for 3–4 minutes; drain and dry in towel; freeze in bags; use within a year.

CAN. Pressure-can, pickled beans can be hot water bath–processed.

Peas—snow, sugar snap, shelling

FRESH. Store in a plastic bag; keep refrigerated; eat within 5 days.

DRY. Blanch shelling peas for 2–3 minutes; dry until wrinkled and hard; will keep for 8 months.

FREEZE. Blanch snow or shelling peas for 2–3 minutes; drain and towel-dry; freeze in bags or containers; best if used within 8 months.

CAN. Pressure-can only.

Carrots

FRESH. Remove tops; store roots in plastic bags; keep refrigerated; will last several weeks.

DRY. Wash and blanch thin slices for 3–4 minutes; dry until tough; use within a year.

FREEZE. Slice, chop, or dice into desired size; blanch 3–4 minutes; drain and towel-dry; freeze in bags; will last up to a year.

CAN. Pressure-can.

Beets

FRESH. Remove tops; store tops and roots separately in plastic bags; keep refrigerated; eat tops within 3 days; roots will last 1–2 weeks.

DRY. Remove tops; cook roots in boiling water until tender; peel and cut into thin slices or small (1/4") cubes; dry until tough and nearly brittle; use within 8 months.

FREEZE. Thoroughly cook beets; peel and cut into cubes or slices; freeze in bags; will keep for 1 year.

CAN. Pressure-can whole beets; pickled beets can be hot water bath–processed.

Squash—summer

FRESH. Keep refrigerated; store in a perforated plastic bag in crisper; eat within 5–7 days.

DRY. Blanch thin slices (1/8" – 1/4") for 3 minutes; dry until leathery; will keep 4 months.

FREEZE. Not recommended unless cooked in another dish; texture of summer squash changes when frozen.

CAN. Not recommended.

Squash—winter (including pumpkins)

FRESH. Store in a cool, dry place. If stem remains attached, will keep for 2–3 months.

DRY. Not recommended.

FREEZE. Bake squash in a 350 degree Fahrenheit oven for 45 minutes or until squash is tender and pack in containers; use within a year.

CAN. Pressure can.

Cucumbers

FRESH. Keep refrigerated; store in a perforated plastic bag in crisper; eat within 7 days.

DRY. Peel, slice thin, and blanch for 1 minute; dry until crisp; will keep 4 months.

FREEZE. Not recommended.

CAN. Pickle cucumbers in a vinegar-based brine and process in hot water bath.

Tomatoes—cherry, slicing, paste

FRESH. Keep uncovered at room temperature; don't refrigerate.

DRY. Halve medium-sized tomatoes, place cut side down on rack; dry until leathery; use within 6 months.

FREEZE. Remove skins; quarter and cook until tender; freeze in containers; use within 4 months. The texture of tomatoes changes when they are frozen, so use frozen tomatoes in cooked dishes only.

CAN. Using a hot water bath, can whole or as sauce.

Broccoli and Cauliflower

FRESH. Keep refrigerated; store in a plastic bag in crisper; eat within 5–7 days.

DRY. Cut into 1/2-inch pieces; blanch for 4 minutes; dry until crisp and brittle; keeps for 1 month.

FREEZE. Cut into 1 1/2-inch pieces; blanch 3–4 minutes; drain and pat dry; freeze in bags; use within 2 months.

CAN. Not recommended, but they can be mixed with other vegetables and pickled.

Radishes

FRESH. Keep refrigerated; remove greens; store in a plastic bag in crisper; eat within 5 days.

DRYING and **FREEZING** are not recommended.

CAN. Not recommended; radishes can be mixed with other vegetables and pickled.

[Recipe] Green Tomato Muffins

Adapted from *KCTS COOKS: Ten Delicious Years—Favorite Recipes*

1 1/2 cup chopped pecans

2 1/4 cups white sugar

1 cup melted butter

3 eggs

2 teaspoons vanilla

3 cups flour

1 teaspoon salt

1 teaspoon baking powder

1 teaspoon cinnamon

1/2 teaspoon nutmeg

2 1/2 cups diced green tomatoes or tomatillos

1 cup currants

1. Preheat oven to 350 degrees F.

2. Place the nuts on a cookie sheet and toast them in the oven for 8 minutes. Cool and chop. Reserve 1/2 cup for topping.

3. Beat sugar, melted butter, eggs, and vanilla until smooth.

4. In a large bowl, stir with a wire whisk to combine all dry ingredients (flour, salt, baking powder, cinnamon, and nutmeg).

5. Add the bowl of dry ingredients to the egg mixture and mix to combine.

6. Stir in 1 cup of pecans, plus tomatoes and currants.

7. Fill muffin cups 3/4 full in two muffin tins. Sprinkle 1/2 cup pecans on top.

8. Bake 40 minutes. Cool on a wire rack.

Yield: 24 muffins

Onions/leeks

FRESH. Store onions in a cool, dark, dry place; don't refrigerate. Keep leeks refrigerated; store in a plastic bag in crisper; and eat within 7–10 days.

DRY. Peel and clean; cut into slices; no need to blanch; dry until brittle; will last 8–12 months.

FREEZE. Slice, chop, or dice onions or leeks into desired size; don't blanch; freeze in bags or containers; use within 8 months.

CAN. Pressure-can baby onions.

Eggplant

FRESH. Keep refrigerated; store in a plastic bag in crisper; eat within 7 days.

DRY. Peel and cut into 1/2-inch strips; blanch for 4 minutes; dry until leathery; keeps for 4 months.

FREEZE. Best when frozen in a prepared dish or lightly sautéed and then frozen. Peel and cut into 1/2-inch slices or cubes; blanch 4–6 minutes; freeze in containers; will keep for 8 months.

CAN. Not recommended. Eggplant can be pickled and canned with other vegetables.

[Recipe]
Leek Tart

Adapted from
Chez Panisse Vegetables

3 medium to large leeks, or about 6–8 cups sliced

3 tablespoons unsalted butter

1/4 teaspoon dried thyme or a few sprigs of fresh thyme

water, stock, or white wine, as needed

Single-crust pastry dough

1 tablespoon flour

1/3 cup soft goat cheese

1 egg, beaten

Salt and pepper to taste

1. Clean leeks, trim off most of the green tops. Slice leeks in half lengthwise then cut into 1/4-inch slices. Rinse well in lots of cold water to remove all sand and dirt.

2. Melt butter in a large skillet. Sauté leeks and thyme over medium heat until tender and almost caramelized—10 minutes. Add a bit of water, stock, or white wine if leeks stick to the pan. Remove from heat.

3. Preheat oven to 400 degrees.

4. Roll out the pastry dough into a 12-inch circle and place it on a baking sheet.

5. Sprinkle the pastry with flour and spread the leeks evenly over the dough to within 1 inch of the edge.

6. Add clumps of goat cheese.

7. Fold the edges of the dough to form a crude shell; brush with egg.

8. Bake on the lowest shelf of the oven until the crust is nicely browned, about 20–30 minutes. Cover the top with a sheet of aluminum foil if it starts to brown too quickly. The tart is done when the bottom of the pastry is nicely browned.

Yield: One 10-inch tart.

Peppers—sweet, hot

FRESH. Keep refrigerated; store in a plastic bag in crisper; eat within 7 days.

DRY. Keep hot peppers whole and dry until brittle; remove seeds and slice sweet peppers into strips; don't blanch; dry until leathery. Both kinds will keep for a year or more.

FREEZE. Remove seeds; slice or dice into desired shape—no need to blanch; freeze in bags; use within 6 months.

CAN. Not recommended for hot peppers; pressure-can sweet peppers.

[Recipe] Green Chile Enchilada Sauce

From Christine Mineart and Graham Golbuff

4–5 mild/medium green chiles (anaheim, poblano, or similar varieties)

¼ cup diced onion

2–3 tomatillos

3 cups heavy cream (or a mix of whole milk and heavy cream)

Salt and pepper to taste

1. Broil green chiles on a baking sheet until the skins are black and blistered. Turn so that both sides are broiled.

2. Carefully peel the skins off the chiles. Remove stems and seeds, and roughly chop the chiles.

3. In a saucepan on medium heat, sauté the onion and tomatillos until they are softened.

4. Add the green chiles and cook a few minutes more. Add the cream and let the mixture heat up, but not boil.

Yield: Makes about 1 quart.

Corn

FRESH. Keep refrigerated in husks; eat as soon as possible.

DRY. Remove husk; let whole cob air-dry until brittle; store on cob or remove kernels and keep in a jar; use within 8 months.

FREEZE. Blanch whole ears for 6–8 minutes; freeze whole ear or cut off kernels; use within a year.

CAN. Pressure-can kernels removed from ears or as creamed corn.

Arugula

FRESH. Keep refrigerated; store in a plastic bag in crisper; eat within 3–5 days.

DRY. Cut into 2-inch pieces; blanch 1 minute; spread not more than a ½-inch thick on trays; dry until crispy; will keep for 2–3 months.

FREEZE. Cut into 2-inch pieces; blanch 1 minute; drain and squeeze out excess water; freeze in bags; will keep for 1 year.

CAN. Not recommended

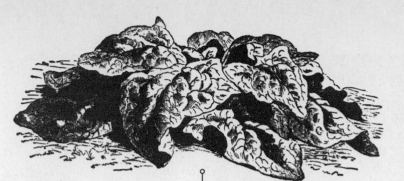

Lettuce

FRESH. Keep refrigerated; store in a plastic bag in crisper; eat within 5–7 days.

DRYING, FREEZING, or **CANNING** Not recommended.

TIP: For quick salads, wash and spin-dry whole lettuce leaves and store in plastic bag in crisper. Grab a few leaves when you need them—will last 5–7 days if lettuce is spun dry before being refrigerated.

Spinach

FRESH. Keep refrigerated; store in a plastic bag in crisper; eat within 5 days.

DRY. Cut into 2-inch pieces; blanch 2 minutes; spread not more than a $1/2$-inch thick on trays; dry until crispy; will keep for 2–3 months.

FREEZE. Cut into 2-inch pieces; blanch 2 minutes; drain and squeeze out excess water; freeze in bags; will keep for 1 year.

CAN. Pressure-can only.

Swiss chard

FRESH. Keep refrigerated; store in a plastic bag in crisper; eat within 5–7 days.

DRY. Cut into 2-inch pieces; blanch 2 minutes; spread not more than a $1/2$-inch thick on trays; dry until crispy; use within 3 months.

FREEZE. Cut into 2-inch pieces; blanch 3 minutes; drain and squeeze out excess water; freeze in bags; will keep for 1 year.

CAN. Pressure-can only.

Kale/collard greens

FRESH. Keep refrigerated; store in a plastic bag in crisper; eat within 5–7 days.

DRY. Cut into 2-inch pieces; blanch 4 minutes; spread not more than a $1/2$-inch thick on trays; dry until crispy; will keep up to 4 months.

FREEZE. Cut into 2-inch pieces; blanch 4 minutes; drain and squeeze out excess water; freeze in bags; will keep for 1 year.

CAN. Pressure-can only.

[Recipe]
Winter Greens with Pomegranate Hazelnut Vinaigrette

From Bev Ham

1 tablespoon olive oil

2 cloves garlic, chopped

1/2 white or yellow onion, finely chopped

2–4 tablespoons balsamic vinegar

1 bunch lacinato or dinosaur kale, rinsed, stemmed. and cut into 1/2-inch strips

1 bunch Swiss chard, rinsed, stemmed and cut into 1/2-inch strips

1/3 cup hazelnuts, toasted and chopped

1 pomegranate, seeded

Salt to taste

1. In a large saucepan, heat oil and garlic over medium high heat and cook, stirring, until golden (2–3 minutes).

2. Add onion and cook until tender (2–3 minutes). Add balsamic vinegar, bring quickly to a simmer, then add kale and Swiss chard, cover pan, and cook until barely wilted (4–6 minutes).

3. Toss with hazelnuts and pomegranate seeds.

4. Season with salt and serve at once.

Makes 4 to 6 servings.

Herbs

Herbs are most commonly used fresh, snipped right from your herb garden. Dry your herbs for use between growing seasons. This is easy and cheaper than buying a little jar at the store. Most herbs are frozen or canned when incorporated in other dishes. Basil, parsley, and cilantro can be puréed, mixed with high-quality olive oil, and frozen in ice cube trays for instant pesto.

Parsley

FRESH. Trim stems and place them in a jar of water, then refrigerate; will keep for 7 days.

DRY. Arrange on rack in a dehydrator or hang the stems in a paper bag to dry in an airy place; crumble dried leaves and store in a jar or bag; will keep for more than 8 months.

FREEZE. Purée with olive oil and a little lemon juice; freeze in ice cube trays and store in bags; use within 6 months.

Sage

FRESH. Refrigerate in a perforated plastic bag; use within 7 days.

DRY. Arrange on a rack in a dehydrator or hang stems in a paper bag to dry in an airy place. Strip dried leaves and store in a jar or bag; will keep for more than 8 months.

Rosemary

FRESH. Refrigerate in a perforated plastic bag; use within 7 days.

DRY. Arrange on a rack in a dehydrator or hang stems in a paper bag to dry in an airy place. Strip dried leaves and store in a jar or bag; will keep for more than 8 months.

Thyme

FRESH. Refrigerate in a perforated plastic bag; use within 7 days.

DRY. Arrange on a rack in a dehydrator or hang stems in a paper bag to dry in an airy place. Strip dried leaves and store in a jar or bag; will keep for more than 8 months.

Oregano

FRESH. Refrigerate in a perforated plastic bag; use within 7 days.

DRY. Arrange on a rack in a dehydrator or hang stems in a paper bag to dry in an airy place. Strip dried leaves and store in a jar or bag; will keep for more than 8 months.

Cilantro

FRESH. Trim stems and place them in a jar of water, then refrigerate; will keep for 5 days.

DRY. Arrange on a rack in a dehydrator or hang stems in a paper bag to dry in an airy place. Crumble dried leaves and store in a jar or bag; will keep for more than 6 months.

FREEZE. Purée with olive oil and a little lemon juice; freeze in ice cube trays and store in bags; use in 6 months.

Basil

FRESH. Put stems in a jar of water and keep at room temperature or refrigerate in a perforated plastic bag; use within 3 days.

DRY. Arrange on a rack in a dehydrator or hang stems in a paper bag to dry in an airy place. Strip dried leaves and store in a jar or bag; will keep for more than 8 months.

FREEZE. Purée with olive oil and a little lemon juice; freeze in ice cube trays and store in bags; use within 6 months.

Chives

FRESH. Put stems in a jar of water and keep them at room temperature or refrigerate in a perforated plastic bag; use within 7 days.

DRY. Cut into small pieces; arrange on a rack in a dehydrator or on newspaper in a dry, airy place; store in a jar or bag; will keep for 6 months.

Flowers

Most edible flowers are used fresh within hours of picking. Some hold up well in cut-flower arrangements, lasting several days before wilting. Other flowers can be used dried in herbal preparations.

Calendula

FRESH. Pluck fresh petals for salad confetti or garnish. Holds up well as a cut flower; keep in fresh water; will last 4–5 days.

DRY. Dry flower heads removed from stems; use the whole flower or remove dried petals.

Bachelor's buttons and Dianthus

FRESH. Pluck fresh petals for salad confetti or garnish. Hold up well as a cut flower; keep in fresh water; will last 4–5 days.

Nasturtiums

FRESH. Use fresh flowers in salads or sandwiches. These delicate flowers will not keep, so eat them as soon as possible.

Sunflowers

FRESH. Wonderful cut flower; will last more than a week.

DRY. Hang flower head in a paper bag or place on a rack in a dry, airy place. Remove seeds and store in a jar or bag.

Borage

FRESH. Use fresh flowers in salads, cakes, or ice cubes. Delicate flowers will not keep; eat as soon as possible.

[RECIPE]
A Salve from Calendula

Infuse calendula flowers in grape seed oil for a homemade salve.

1. Fill a small jar with blossoms and pour oil over them.

2. Cover and let sit for a day on the counter and then refrigerate for a week.

3. Strain the oil and use on sunburns, scrapes, and rashes. Will keep in the fridge for up to 2 months

Tree Fruits

The fruits of your labors are some of the most versatile foods you will harvest from your city farm. Fruit can be eaten fresh, dried, frozen in pies, and canned. Keep fruit slices or leather from turning brown by dipping them briefly in dilute lemon water (1 tsp lemon juice to 2 quarts water) before dehydrating. Freeze fruit as slices, or in sweet liquid. Hot water bath–can fruit as sauce, butter, jam, or in pieces in light syrup.

Apples

FRESH.

Refrigerate loose in crisper; will keep 1 month.

DRY. Peel, core, and thinly slice; dip into lemon water; dry until leathery; will keep up to 2 years.

FREEZE. Dip peeled slices in lemon water and pack in containers; prepare sauce, butter, or jam and freeze in containers; eat within 6 months.

CAN. Hot water bath-can prepared as slices, sauce, or butter.

Cherries

FRESH. Keep refrigerated; store in a perforated plastic bag in crisper; eat within 5 days.

DRY. Pit and half cherries; dip in lemon water; dry until leathery; use within a year.

FREEZE. Wash, pit, and dip in lemon water; freeze in bags or containers; use within 6 months.

CAN. Hot water bath–can whole in syrup.

Pears (European and Asian)

FRESH. Let ripen on counter at room temperature, then refrigerate in plastic; eat within 5 days.

DRY. Peel, core and thinly slice; dip into lemon water; dry until leathery; will keep 6–9 months.

FREEZE. Not recommended; pears lose color, texture, and taste when frozen.

CAN. Hot water bath–can prepared as slices, sauce, or butter.

[Recipe]
Ginger Pears

From Laura Matter

8–10 lbs ripe pears

**1 teaspoon ascorbic acid or
1 teaspoon lemon juice**

**1–2 quarts of water to dilute
acid or lemon juice**

FOR SYRUP:

2 cups sugar

4 cups water

3 teaspoons lemon juice

**$1/3$ cup candied ginger,
chopped small**

1. Peel, core, and slice 8 to 10 pounds of ripe pears into quarters.

2. Place into cold water bath with 1 teaspoon ascorbic acid dissolved in one pint of water or 1 teaspoon lemon juice diluted in 2 quarts of water—to prevent discoloration.

3. Soak for no more than 20 minutes; rinse and drain.

4. Prepare light syrup with ginger and lemon juice: 2 cups sugar to 4 cups water and 3 teaspoons lemon juice and $1/3$ cup chopped candied ginger. Bring it to a boil and keep on a low boil.

5. Blanch drained pears for 5 minutes in hot apple juice or in your ginger lemon syrup.

6. Pack hot pear slices into hot jars leaving $1/2$ inch of headspace. Pour boiling syrup over pears, and remove air bubbles with a chopstick or small rubber spatula. Adjust lids.

7. Process in a hot water bath according to the size of the jar and your altitude. Typically: Pints for 20–25 minutes and quarts for 25–30 minutes.

8. Remove your jars from the water bath with a jar lifter; let them air-cool on a dish towel. Allow jars to seal. Let stand for 12 hours. Test lids for seal. Refrigerate any that do not seal.

Makes 6 to 8 pints.

TIP Substitution: You can substitute a cinnamon stick for the ginger. Remove sticks from the syrup before canning.

Plums

FRESH. Let ripen on counter, then refrigerate; store in a plastic bag in crisper; eat within 5 days.

DRY. Pit and cut into 1/2-inch strips; dry until leathery; will keep for a year or more.

FREEZE. Remove skins, pit and halve (purée if desired); freeze in light syrup in containers; will keep 6 months.

CAN. Hot water bath—can prepared as slices, sauce, or jam.

Peaches/nectarines

FRESH. Let ripen on counter; store in a plastic bag in crisper; eat within 5 days.

DRY. Pit and thinly slice; dip into lemon water; dry until leathery; use within a year.

FREEZE. Remove skins, pit and halve. These fruits brown easily and lose texture when frozen, so dip slices or halves in lemon water, mix in light syrup, and pack in freezerproof containers; will keep 6 months.

CAN. Hot water bath—can prepared as slices or jam.

Apricots

FRESH. Let ripen on counter, then refrigerate; store in a plastic bag in crisper; eat within 5 days.

DRY. Pit and thinly slice; dip into lemon water; dry until leathery; will keep for more than a year.

FREEZE. Remove skins; pit and cut into slices or halves; dip in lemon water; freeze in syrup; add a few pits to each container for flavor; use within 6 months.

CAN. Hot water bath—can prepared as slices or jam.

Shrub Fruit

Figs

FRESH. Keep refrigerated; store in a paper bag or in a perforated plastic bag; eat within 3–5 days.

DRY. Cut in half; dry until leathery; use within a year.

FREEZE. Prepare whole or halved in light syrup with lemon juice in containers; eat within 4 months.

CAN. Hot water bath—can prepared as halves, sauce, or jam.

Elderberries

FRESH. Do not store well; process the same day they are harvested. Berries should not be eaten fresh, skins and seeds are mildly toxic.

DRY. Remove berries from stem and dry until brittle; will keep for more than a year.

FREEZE. For syrup, cook berries with stems removed until they are soft; strain liquid, sweeten to taste, and pack in containers; will keep for 4–6 months.

CAN. Hot water bath—can prepared as juice or syrup.

Blueberries

FRESH. Keep refrigerated; store loosely covered in a bowl or berry basket; eat within 3 days.

DRY. Remove berries from stem; dry until tough and leathery; use within 8 months.

FREEZE. Arrange on trays in the freezer; store in bags or containers after frozen; will keep for 6 months.

CAN. Hot water bath—can prepared as jam or jelly.

Currants

FRESH. Keep refrigerated; store in a plastic bag or shallow container; eat within 3–5 days.

DRY. Remove berries from stem; dry until tough and leathery; eat within 8 months.

FREEZE. Arrange on trays in the freezer; store in bags or containers after frozen; will keep for 6 months.

CAN. Hot water bath—can prepared as juice or jam.

Strawberries

FRESH. Keep refrigerated; store loosely covered in shallow bowl or berry basket; eat within 2–3 days.

DRY. Remove stems; halve and dip in lemon water; dry until leathery; will keep for 6 months.

FREEZE. Arrange on trays or mix with syrup, then place in the freezer; pack frozen strawberries in bags or containers; will keep for 6 months.

CAN. Hot water bath-can prepared as jam or jelly.

Cane Fruit

Raspberries

FRESH. Keep refrigerated; store loosely covered in a shallow bowl or berry basket; eat within 2–3 days.

DRY. Not recommended.

FREEZE. Freeze whole on trays or mix with syrup; store in bags or containers; will keep for 6 months.

CAN. Hot water bath—can prepared as jam, jelly, or syrup.

Blackberries

FRESH. Keep refrigerated; store loosely covered in a shallow bowl or berry basket; eat within 2–3 days.

DRY. Not recommended.

FREEZE. Freeze whole on trays or mix with syrup; pack in bags or containers; will keep for 6 months.

CAN. Hot water bath—can prepared as jam, jelly or syrup.

Fruit Grown on Vines

Grapes

FRESH. Refrigerate in a perforated plastic bag in crisper; eat within 5-7 days.

DRY. Remove grapes from stem; dry until tough and leathery; use within 8 months.

FREEZE. Sweetness is intensified through freezing; freeze on trays; store in bags or containers; will keep for 6 months.

CAN. Hot water bath—can whole grapes in light syrup or prepare as jam or juice.

Summer All Year Round

Get the most out of your city farm by extending your growing season and planting cool-season crops again in July and August for a harvest of fall greens, peas, or beets. Keep eating from your garden all year round by freezing, drying, or canning your bounty. Preserving food for use all winter not only makes economic sense, but also brings a little of summer's glory back to the kitchen, even in the dead of winter.

Profile: THE FOOD PROJECT

Some say that there is nothing more important than growing food. The folks at The Food Project (TFP) in Boston agree, but they also believe that there is nothing more important than giving young people responsibility and teaching them life skills. In this case, these goals are one and the same.

Since 1991, The Food Project has worked with nearly 100 teenagers and thousands of volunteers to farm 37 acres in eastern Massachusetts in its Summer Youth Program. Most of these young people have little or no experience growing food. Many of them have never even been to a farm. Teens hired by TFP learn to grow food; more importantly, they get to see firsthand the importance of food for their communities. During the 6½-week program, young people work full time farming, harvesting, and selling produce at TFP's numerous farmers' markets. About 20 percent of what they grow is distributed through local food pantries and meal programs.

THE FOOD PROJECT
Youth • Growing • Together

The Food Project's Academic Year Program hires summer graduates to help bring in the rest of the harvest. The DIRT crew (stands for Dynamic, Intelligent, Responsible Teenagers) works afternoons and Saturdays to finish off the growing season. Each year The Food Project staff, interns, volunteers, and youth grow and distribute nearly 15,000 pounds of food!

From providing resources to hands-on experiences, The Food Project sets up today's youth for success now and in the future by emphasizing the connection among people, gardens, and food—highlighting the fact that city farming means real food for communities.

LEARN MORE:
thefoodproject.org

URBAN FARM ANIMALS

AFTER A FEW YEARS OF growing vegetables, many city farmers expand their food production to include animals in their operation. Keeping animals in town isn't a new trend—folks have quietly kept a few chickens or rabbits in their backyards for centuries. However, land use ordinances are changing to allow a wide variety of farm animals to be kept within city limits. Urban farmers are stretching the limits of what can be grown on a city lot—keeping beehives, miniature goats, even flocks of ducks. In this chapter, you will learn what it takes to keep chickens, ducks, rabbits, goats, and bees on your city farm.

Integrating Livestock into Your Urban Farm

Why keep animals?

There are a number of reasons to keep animals on your city farm. They can be raised to produce food, such as eggs, milk, meat, and honey. Some animals are prized for the valuable fiber that can be harvested from them, such as angora wool, mohair, and cashmere. Farm animals can also enhance your garden. Chickens, ducks, rabbits, and goats consume garden debris and contribute manure to the garden.

Bees pollinate flowers and increase fruit production. Additionally, urban farm animals are often adorable and can provide you with hours of entertainment.

Are farm animals for you?

Before you leap into animal husbandry, do your homework. Learn all about the animals before you even think about buying babies. Research housing and care needs, feeding, raising young, and harvesting animal products. Chickens, ducks, goats, and rabbits will require daily care and feeding. Bees require weekly maintenance and monitoring during the growing season. Also, become familiar with persistent or serious diseases and how they are managed.

Decide what your goals are for keeping livestock. Are the animals pets, sources of fiber, or sources of food? What breeds fit your space? Check with land use codes to make sure it is permissible for you to keep livestock on your property.

Pick up a good manual on raising the animal(s) you're interested in. Take classes and talk with others who raise farm animals in the city. Better yet, talk with someone who no longer keeps livestock and learn about the drawbacks or difficulties associated with the animal(s). Find the experts and enthusiasts in your area— local growers' associations and your cooperative extension service are excellent sources of information and can help in locating animals and supplies.

Participate in urban livestock activities before setting up your operation. You may be able to apprentice with a beekeeper or another city farmer. Try your hand at milking a dairy goat or help with butchering a few times before you go into dairy or meat production.

Can I save money?

Most farm animals aren't too expensive to purchase but the equipment, housing, and feed can certainly add up. Unless you are setting up a larger operation, you won't save much on, say, eggs or honey. In a few years, after you have recouped your initial investment, it may still be a wash if you figure in the time and money spent caring for the creatures and fetching feed. Most city farmers who raise animals aren't as focused on the bottom line—they keep animals because they are fun, the eggs are beyond compare, and they get to play farmer.

Pets or meat?

Many people keep animals as a productive novelty, such as a few laying hens or ducks that are part of the family and as an entertaining feature on a city farm. These are considered no-kill livestock. These animals will live out their natural life span until they succumb to predators, illness, or old age.

Some city farmers raise animals as a local source of meat. Chickens, ducks, and rabbits are efficient at converting feed to protein and reach market weight in two months. In hard times, this is an attractive idea. Be sure to do some butchering or dressing of the animals before you start keeping them for meat, or you may end up with a ranch full of livestock that you simply can't kill. Above all, be sure to use exceptional sanitation in handling meat.

Whatever your goals for keeping farm animals in the city, here are the basic care, housing, and food requirements for some common urban livestock.

Livestock: The Basics

If you are new to keeping livestock, chickens are a great place to start. They are inexpensive and start laying eggs in four to five months. After your coop is constructed, they require only a small amount of daily care.

BENEFITS

Chickens can be raised for meat, eggs, and high-quality manure. Though some people raise a few chickens each year for meat, most city farmers keep chickens for eggs, manure, and endless entertainment.

Chickens are great cultivators; they love to dig up the garden; aerate the soil; eat insects, slugs, and worms; and add fertilizer. You can construct a chicken tractor, a movable chicken coop without a floor, and place it where you want the hens to dig.

Chicken poop is great for the garden and hens are great recyclers. They will eat weeds, old plants from the garden, and kitchen scraps, which become high-quality manure that can be used to grow more food.

Plus, fresh, homegrown eggs are superior to even organic eggs sold at the supermarket. They have deep, amber yolks that stand up tall in your skillet and the whites are viscous and not runny. They taste fresh and rich.

BASIC CARE

Keeping chickens requires one to three quick visits each day. Every morning birds will need to be fed and let out of the henhouse. In the evening, eggs are collected and more feed scattered. You may make one more trip to close up the henhouse if your chicken run isn't totally raccoon-proof. If you have a well-constructed, raccoon-proof enclosure, you may only need to visit your hens once a day for egg collection and feeding. Hens will lay without a rooster—the eggs will just not be fertilized.

CHICKENS

Each hen lays between 250 and 288 eggs per year, but production decreases after two years. Hens make quite a ruckus each time one of the flock lays an egg.

Chickens eat different feed depending how old they are. Chicks are typically fed a starter mix for the first 6 weeks, and then are switched to a grower mix until they are 20 weeks old (though they will do just fine on layer mix after the starter mix runs out). After chickens are 5 months old, they can be fed a layer mix. Chicken feed comes in either a crumble (or mash) or as pellets. I've found there is much less wasted with pellets.

A small flock of three or four chickens will eat 1–2 pints of feed each day and still beg for more. Supplement feed with kitchen scraps that are easy for chickens to eat, such as rice or pasta, as well as eggshells for needed calcium. Toss in some dandelion or other greens for rich eggs with dark amber yolks. Don't feed chickens large, hard food scraps, such as onion skins or broccoli; these won't be eaten and will attract rodents.

WINTER CARE

Provide plenty of clean straw in nests, and close up your henhouse at night against cold weather. You'll need to break the layer of ice to make sure there is always water available. In extremely cold climates, insulate your henhouse or add a 60-watt bulb for heat.

HOUSING

Chickens need protection from harsh weather and to be kept in a well-constructed, raccoon-proof enclosure. The biggest initial cost in keeping chickens is your coop. City coops can range from do-it-yourself models that cost a couple hundred bucks for materials to luxurious chicken mansions that cost over a thousand dollars and come ready to assemble.

Your coop will include a henhouse where the chickens will spend the night and a run area where the chickens will spend the day. Within the henhouse is a nesting box, where eggs are laid. The henhouse and run are referred to as the coop. At a minimum, your coop will need to allow 3–4 square feet of floor space per bird. Make sure your henhouse has good ventilation without drafts with 1–2 square feet of space for each bird. Many nesting boxes are mounted to the henhouse with access from outside the coop for easy egg collection. The henhouse door should be wide, so that it is easy for you to clean out old bedding and manure.

Make sure your coop has adequate lighting, because birds need sunlight to stimulate egg production. Wrap your coop in poultry netting or hardware cloth so birds are safe from predators. Bury wire 12 inches deep or bend a foot of netting outward away from the coop and bury it under 4–5 inches of soil around the edges of your enclosure. This will deter predators from digging underneath your

Typical Chicken Coop

enclosure. Build a perch in the henhouse and coop so the chickens can roost. To stay healthy, chickens need to take dust baths, so make sure there is always a dry dirt area for your hens.

Decide early on if your hens will be kept confined or will be allowed to roam around the yard. If your hens have tasted a bit of freedom, they will not be satisfied with their enclosure, no matter how big or luxurious it is. They will complain loudly each time you walk by the coop.

Chickens love space and want nothing more than to explore, leaving in their wake some mighty fine fertilizer. They are skilled hunters of ground-dwelling insects, slugs, and grubs. They love to eat errant clover and dandelions in the lawn, but they will wreck your garden if left to wander unchecked. If your chickens range freely, you will need to fence off your garden beds.

RAISING CHICKS

Most people start with day-old chicks or peeps purchased from a feed store or ordered by mail. Peeps need to be kept warm until their feathers have grown in. Raise them in the house or in a heated garage for the first

couple of months. A large box with wood shaving litter and a 60-watt bulb for added warmth will work well. Use a large plastic trunk (18" x 40") rather than a cardboard box—it will be easier to clean and large enough for the first few weeks. After a month, move your chicks to a makeshift pen and keep the temperature at 75 degrees Fahrenheit. Chicks need to be kept warm as they grow. The temperature in their box should be 95 degrees for the first week and then reduced by 5 degrees each week until the ambient temperature is reached or the weather is warm enough for the birds to go outside. Adjust the temperature by raising or lowering the light in the box. Use a thermometer to monitor the temperature or watch the chicks. If it is too cold, they will huddle under the light and shiver. If it is too hot, they will stay at the end of the box farthest from the lamp. If the box is too large

BREEDS FOR CITY FARMS

Day-old baby chicks can be purchased in early spring at a feed store or via mail order. Docile breeds that are sturdy and reliable layers for city farms include **Black Australorp, Buff Orpington, Plymouth Barred Rock, Rhode Island Red, Sexlinks,** and **Silver-laced Wyandot**. Typical breeds are inexpensive ($5–10 dollars apiece). Fancier breeds and purebred chickens will cost more.

and cold, put a cover on the end away from the light; this will help keep the space warm. Make sure to keep peeps safe from other family pets.

Water and chick starter mix, dispensed in a feeder, should be consistently available to peeps. Chickens also need grit for their gizzards, where food is ground up. Give your peeps a good start by adding a handful of dirt or some castings from the wormbin to their feed when they are a couple weeks old. Chicks benefit from all the creatures in worm compost and will grow to be strong, healthy, and resistant to disease on a diet laced with worm compost.

Hens start laying eggs when they are between 20 and 24 weeks old. Egg production is spotty for the first month, but by the time they are 30 weeks old, each hen will lay 2 eggs every three days. A hen will typically lay about 20 dozen eggs her first year; 16–18 dozen eggs during her second year. Production drops more rapidly in years 3 and 4. By the time a hen is five years old, she may only lay 1 egg a week. Hens usually die of old age before they turn six.

At some point the cost of feed may outweigh the value of eggs and manure produced. Most commercial operations don't keep layers that are more than two years old. On some farms an old chicken is headed for the stew pot, but many city chickens are kept until they die naturally. A good livestock manual will show you how to butcher, pluck, and dress poultry—the process is messy and rather labor intensive but not complicated.

Chickens molt their feathers each year in the fall. Molting takes about two months and during this time egg production will drop or the hen will stop laying completely. She needs all her energy to make new feathers. Egg production will also slow or stop in winter because of shortened daylight. You can add a 40-watt bulb in the henhouse to keep them laying through the winter. Many city farmers like to give the "girls" a rest in the winter and respect the chickens' natural cycle.

HARVESTING EGGS, MEAT, AND POOP

Eggs should be collected daily. Brush them off to remove persistent manure. Keep straw in the nesting box clean and eggs will be cleaner. Wash eggs in warm water when you are ready

A hen will typically lay about 20 dozen eggs her first year.

to eat them. Raw eggs kept in an egg carton on the bottom shelf of the refrigerator will be good for four weeks. Keep track of egg freshness by writing the date in pencil on the end of each egg.

Sometimes you'll find an egg in the corner of the coop and it is impossible to know how old it is. Test for freshness by dropping the egg in a bowl of cold water. If the egg is fresh, it will sink. There is less air space in fresh eggs; as moisture evaporates inside the egg, there is more space, so an old egg will float. If an egg floats, it is too old to be eaten.

COMMON PROBLEMS

Noise, dust, flies, and odor are the major drawbacks of keeping city chickens. They kick up dust and make a lot of noise when they lay eggs. Some hens will imitate a rooster and herald the morning sun in a delicate cock-a-doodle-do. Unless you are fastidious about poop scooping, you will have flies and odors. Any chicken that might escape from the coop will tear up the garden. Hens will eventually stop laying, which means you will either need to run an old age home for these hens or butcher them for the stew pot.

Roosters aren't needed for egg production and aren't the best choices for city farming. Many cities that allow chickens discourage or prohibit roosters. If you should happen to get a rooster by accident, contact your cooperative extension service to find a farm where the rooster can be relocated.

Chickens can't see at night and are extremely vulnerable to predators. Your enclosure should be able to withstand assaults by dogs, hawks, falcons, owls, raccoons, skunks, foxes, weasels, and opossums. Cats are not normally a threat, but rats like to eat chicken food and sneak into the henhouse at night to nibble on your hen's feet—ick!

There are a few diseases that affect city chickens. Most attack baby chicks (though there are some that afflict adult hens) and result in sudden death. Monitor your chickens for unusual behavior and isolate any sick chickens immediately. Be sure to always wash your hands before and after handling a sick animal.

ADDING NEW CHICKENS TO YOUR FLOCK

After a couple of years, your chickens will produce fewer eggs or you may have lost a few hens to predators or disease. Whatever the reason, you may want to add some new chickens to your flock. The pecking order is a real and sometimes ugly thing. Baby chicks put in with full-grown hens will be pecked to death. Even adolescent chickens will be picked on.

Raise the new chickens separately until they are about full size, but before their waddles and comb are fully grown. Split your coop in half—using plastic mesh netting or a wall of chicken wire. Put a nesting box, water, and food for the new hens on their side. This will give the older chickens a chance to get adjusted to the new birds. After a few weeks, when the new hens have their full waddles and comb, lift the wall and see what happens.

DUCKS

 ucks are kept by city farmers for eggs, meat, and to control pests in the garden. Caring for a flock of ducks is similar to keeping chickens, with a few differences.

BENEFITS

Egg-laying duck breeds can provide a steady source of eggs. Duck eggs are a food staple the world over, but they don't taste like chicken eggs and many people can't adjust to the stronger, gamier flavor. They can be leathery when fried, but are prized for baking. To see if you like the taste, sample a few dozen duck eggs before you make the leap into duck husbandry.

Duck droppings are watery and hard to collect. However, bedding from your duck shelter will make a good addition to your compost bin or can be used as mulch. Ducks are great for controlling pests in your garden. They love to eat slugs, snails, and other ground-dwelling insects. They will also eat your plants, but will not dig and tear things up, the way chickens will.

BASIC CARE

Ducks are hardier than chickens, but need more space. They don't need a pond (though they would love it), but they do need water for washing their heads. Make water constantly available in a wide, low trough or bucket. They will need protection from harsh weather, clean housing, and quality food. Ducks should be confined in their shelter at night to protect them from predators. Make sure there is adequate ventilation, but that their shelter is not drafty.

Feed your ducks a waterfowl mix (chicken feed can be substituted in a pinch). Ducks don't peck or eat food off the ground, so food should be served in a heavy, shallow bowl. They will also eat kitchen scraps—experiment to find out what they like best.

WINTER CARE

Ducks won't need extra care during the winter. Make sure there is plenty of dry bedding in their shelter and in their yard. Since ducks stay on the ground (they don't roost, like chickens), if their housing area is too wet with droppings, their feet can freeze to the ground.

HOUSING

The main cost of keeping ducks is the shelter and duck yard. Even if you are an expert scavenger, it will cost a couple hundred dollars for the shelter and run; more if you need to fence a larger area. The duck shelter should provide 4–5 square feet per bird, plus a large yard or run. Even meat breeds will lay a few eggs (less than 100) each year, so you will need to provide a nesting box in your shelter. Unlike chickens, nesting boxes for ducks don't need a top, so an old crate works just fine.

Ducks make a nonstop supply of watery poop and can turn a lawn area into mud quickly. A 5 x 8 foot shelter plus a 400-square-foot duck yard would be adequate for as many as five ducks. As with chicken coops, your duck shelter and yard must be predator- and rat-proof.

A pond isn't needed, but if you provide one, make sure the water stays clean and isn't stagnant. If you have a large lot, ducks can wander and forage during the day. If your ducks will free-range during the day while you're at work, run an electric wire around the area to keep ducks in and predators out. Some farmers clip ducks' wings to render them flightless, but this isn't necessary, since they can be easily trained to come for food and will be unlikely to wander off.

RAISING BABIES

Raise ducklings in a large plastic box or a 4-foot kiddie pool with 3-foot cardboard walls surrounding it. Keep their environment warm, as you would with baby chicks. Start at 90 degrees Fahrenheit and reduce the temperature by about 10 degrees every seven to 10 days. When ducklings are 4 weeks old and fully feathered, they can be turned outside.

HARVESTING EGGS AND MEAT

Eggs should be collected daily, dated, and kept refrigerated. They will last 4 weeks if kept in a carton on the bottom shelf of the fridge. Breeds raised for meat reach market weight in 7 to 10 weeks. Butcher ducks as

Typical Duck Enclosure

you would chickens and either dry-pluck or scald them to remove feathers. A good livestock manual will show you how to butcher, pluck, and dress poultry—the process is messy and rather labor intensive but not complicated.

COMMON PROBLEMS

Duck poop smells and duck yards can become a muddy mess. If you are keeping ducks in a smaller run, use 6–8 inches of pea gravel as the floor. Poop can easily be hosed down into the gravel.

Ducks quack incessantly. Males of most breeds are silent, but female ducks always have something to say. If noise is a concern, the Muscovy is considered a "quackless" breed: Females are silent and the males make a quiet hissing sound.

There are no major diseases associated with small flocks of ducks; however, you will need to isolate any sick ducks. Keep their housing clean, provide quality food, and monitor ducks for any odd behavior.

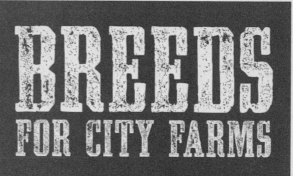

BREEDS
FOR CITY FARMS

Baby ducklings are inexpensive (about $5–10 apiece) but they may be more difficult to find than chicks. Look for duck breeders in your area or purchase by mail from a reputable hatchery.

Ducks are classified as egg or meat breeds; there is no such thing as a dual-purpose duck.

Egg-laying breeds are small, weighing only 4 pounds, and are quite well suited for a city farm. Dependable laying breeds, such as **Khaki Campbell** and **Indian Runner**, will lay 200–300 eggs per year. Meat breeds are much larger— **Muscovy**, **White Pekin** and **Aylesbury** tip the scale at 8–10 pounds—and need more space.

RABBITS

Rabbits may be an excellent addition to your city farm. They are quiet, easy to care for, make affable pets, and consume weeds, such as blackberry and chickweed. They also make great manure for your garden. Rabbits don't require much space or much care.

BENEFITS

Rabbits are the ultimate multipurpose small livestock, perfect for small city farms. Rabbits can be raised for wool, fur, compost, and meat. Traditionally raised for meat, rabbits are great family pets and when raised for wool and manure are no-kill livestock. Angora rabbits produce top-shelf wool fiber. Many farmers enjoy showing purebred rabbits.

BASIC CARE

Rabbits require very little care. They need to be fed each evening and checked daily for general health. Water must always be available. Keep the floor of their hutch clean and dry. Mold will grow in moist hay or feces, which can irritate your rabbit's feet and is a perfect place for disease to start.

Cages should be cleaned on a regular basis. Take rabbits out of their cages and use a wire brush to remove any manure or hair stuck to the wire. Wash cages with a mild bleach solution (1 Tbsp bleach in 1 gallon water). Rinse well with water and let the cage dry completely before returning the rabbits.

Feed rabbits a commercial pellet blend that provides 100 percent nutrition and supplement that with loose alfalfa hay. Organic rabbit pellets are often competitively priced and can be purchased in 3–50 pound bags. As with other livestock, keep feed clean, dry, and away from rodents. A small galvanized can with a tight-fitting lid works perfectly to store feed.

Provide only as much feed as your rabbits can finish in a day—about ¼–¾ cup, depending on the breed. Food and water can be given in small bowls or cans, but these spill easily and are fouled quickly by feces and hay. Instead, use

WINTER CARE

As long as your rabbits stay dry and are in a protected place that isn't drafty, they should be fine in the winter. You can insulate the roof of your rabbit hutch to better regulate extremes of heat and cold. In the frozen North, rabbit hutches may be moved into an unheated garage or shed and a heating cable for the water will be needed to prevent water from turning to ice. Monitor rabbits and provide supplemental heat if needed.

a water bottle with a feeder tip and a pellet dispenser that is mounted on the outside of the hutch. For each hutch, provide a salt spool, which provides minerals, and a block of clean wood for the rabbit to chew to maintain good oral health.

HOUSING

Rabbits should be kept in a sheltered location, out of direct sun, wind, and rain. Make sure rabbit hutches are in a protected place, such as in a garage, under a carport, or against a shed. Hutches can be constructed out of wooden frames and wire fencing as a do-it-yourself project. Cover hutches with a waterproof roof that has a generous overhang, especially if you live in a rainy climate. Cages should have good access. Doors should open easily and be wide enough to clean out the cage.

The cage floor should be metal or wire. Hutches can rest on concrete or dirt or be raised above the ground. Wooden floors are not recommended; they will absorb urine, creating foul odors and unsanitary conditions. You will need a water dispenser, a feed hopper, a manger for hay, and a nesting box if you plan to breed your rabbit(s).

Each hutch should be large enough to allow your rabbit to turn around. Hutches are typically

Typical Rabbit Hutch

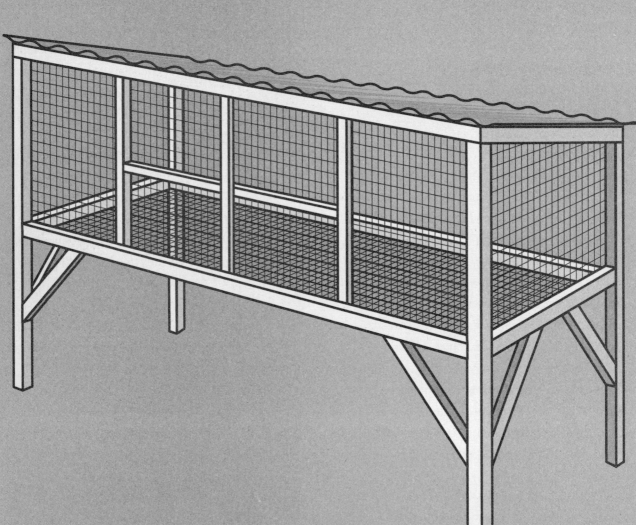

30 x 36 inches for medium-weight or smaller breeds—this will provide plenty of space for a doe and her litter. A hutch for a giant breed should be 30 x 48 inches. Simple all-metal cages start at $60, while deluxe wood and wire hutches, with attached run, cost $300 or more.

If you are building your own hutches, use welded wire fencing that has a 1 x 2 inch grid. Don't use hardware cloth for the floor of your hutch. Hardware cloth is made out of woven galvanized wire. This rough and sharp surface can injure the rabbit's feet. Feces and fur stick more easily stick to hardware cloth, causing unsanitary conditions that invite disease.

Rabbits can also be kept indoors and trained to use a litter box. Choose a biodegradable litter that has not been treated with an antifungal agent. Avoid cedar shavings; they break down slowly in the compost pile and cedar oil is caustic and may irritate sensitive rabbit skin.

RAISING BABIES

If you buy young rabbits, they will be weaned and ready to live on their own. Provide them with food (1 oz. daily for each pound of rabbit) and make sure water is always available. If you are breeding rabbits, the doe will nurse her kits and they will eat rabbit feed as their eyes open and they move about the hutch. A doe and her litter will eat about 100 pounds of feed in eight weeks. There should be enough room in a 30" x 36" hutch to keep kits with the doe until it is time for butchering.

HARVESTING MEAT, WOOL, AND POOP

A good rabbit husbandry manual will show you how to butcher rabbits for meat. The process is simple compared to scalding and plucking chickens or ducks. Still, butchering isn't for everyone—you may be able to find another rabbit farmer or a specialty

BREEDS
FOR CITY FARMS

There are many different breeds of rabbits raised around the world. Rabbit breeds are distinguished by size and weight. There are four classifications: giant, medium, small, and dwarf. The breeds most commonly raised are medium weight—**Californians** and **New Zealand Whites**—and are grown for meat production because they convert food to protein efficiently and reach butchering weight quickly. There are several angora breeds that produce highly valued wool fiber. **Angoras** are a naturally docile breed and make great indoor pets.

Baby rabbits are inexpensive (less than $25 apiece) but purebred breeds can run a couple hundred dollars each. Always buy rabbits from a reputable breeder. Spaying or neutering a rabbit can cost $50–200. Check to see whether there is a rescue agency or veterinarian in your area who will spay bunnies.

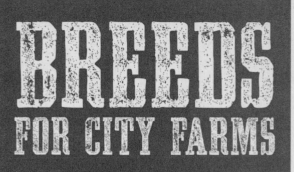

butcher who would clean, dress, and wrap your rabbit meat. You can also use the fur pelts from meat rabbits for handmade crafts.

Rabbits grown for wool are plucked or sheared to coincide with molting. Angoras molt four times per year. As the new coat comes in, the old fibers are loose and can be brushed and plucked from the rabbit's body with your fingers. You can also cut the wool with sharp fabric shears—be careful, however; the wool fibers dull blades quickly. Angoras can produce nearly a pound of fiber each year. Angora wool is seven times warmer than sheep's wool. The market for angora fiber is strong—raw fiber can be sold for spinning or can be spun and sold by the skein.

Rabbit poop is a valuable resource for city farms. It is relatively low in nitrogen and can be used immediately in the garden without burning plants. Many industrious city farmers use rabbit droppings as a food source for growing worms that are sold for fishing or composting.

COMMON PROBLEMS

Diseases don't pose much of a problem for city rabbits as long as they are fed well and kept in clean, dry hutches. Monitor them daily to observe the health of your rabbits. Check for sores on their feet, as this is a place where diseases often start. Isolate sick rabbits. Wash your hands before and after handling each rabbit.

Rabbits frighten easily. In a cage, they can't run and sometimes injure themselves or die of fright. Keep cages in an area that is protected from dogs and other animals that might be perceived as a threat. Hutches that are raised 3 or 4 feet above the ground are at a convenient height for tending and better protect rabbits from being frightened should a stray dog run into your yard.

Rats can eat rabbit food or kill baby rabbits. Give rabbits only what can be eaten each day, keep cages clean, and monitor the hutch for rodent activity. For extra protection, add an additional outer layer of ¼-inch hardware cloth to the sides of the hutch. If you are lining the bottom, make sure there is a space between the hutch floor and the hardware cloth so that rabbit feet don't come in contact with sharp wire mesh.

RABBITS FOR MEAT

Rabbit is becoming a popular local protein source. Rabbit meat is delicate, delicious, and nutritious. It is higher in protein and lower in fat, calories, and cholesterol than chicken, beef, goat, duck, turkey, or pork. Rabbits grow quickly to fryer size and are easy to butcher and dress.

Rabbits convert food to protein more efficiently than chickens or ducks. In a space as small as 5 x 3 feet, you could produce enough meat to feed the family. You'll need space for two hutches—one for a buck and one for a doe. A doe can have five litters of seven kits each year. Young rabbits reach market weight (3–4 pounds) at about eight weeks old. Enterprising city farmers market fryers to local chefs or restaurants to offset the costs of feed.

GOATS

 Keeping goats is a major commitment that requires a lot of time. Goats need plenty of space and dairy goats must be milked twice a day. Make sure you know what you are getting into before purchasing goats. Read a good manual and talk to other people in your city who raise goats.

BENEFITS

Goats are a highly entertaining and mischievous addition to your city farm. Goats are very social and those that have been well socialized by humans can be like a family dog. They are friendly and calm, with great personalities. As long as you aren't trying to get them to do something they don't want to do, they are cordial and loving. During chivalric times, goats were housed with horses to keep them calm. The expression "Someone's got your goat" comes from the practice of opposing armies sneaking into the enemy camp and stealing their goats. The abandoned horses would then panic.

Some people raise goats and hire them out as four-footed weeding machines. Cities hire these goatherds to clear areas that are inaccessible to people and machinery. Keeping a weeding herd of goats requires more space than most city properties allow.

Most city farmers keep goats for milk. Miniature and dwarf dairy goats are becoming a popular urban livestock choice. Dairy goats produce fabulous milk.

Fresh goat milk is far superior to the goat milk sold in stores and is virtually indistinguishable from cow's milk. Goat milk is easier to digest and can be consumed fresh or made into cheese and butter.

BASIC CARE

Dairy goats are high-maintenance animals. They will need a shed for shelter, daily food and water, and a large yard for passing the day. They must be milked twice a day, in the morning and evening. Expect to spend half an hour each morning and each evening milking, feeding, and cleaning up the goat pen. Add to this another hour each week on other goatherding chores. Milking goats is easy, but it takes a week or two of practice to become a proficient milker. You will need to arrange for a competent goat sitter if you plan to take a vacation.

Goats are picky eaters, though they will eat a variety of plants, shrubs, and woody material. Feed them high-quality alfalfa hay and a grain ration. Goats will not eat from the ground, so make hay available from a raised manger. Provide at least 1 foot of manger space for each adult goat in your herd. Two dairy goats will eat 200 pounds of alfalfa pellets and half a bale of hay each month, which costs about $75. Lactating does will require a few cups of grain each day. Make water available at all times.

HOUSING

Provide goats with protection from sun, wind, rain, and snow. Your goats will need a simple shed so they can get out of severe weather. Goats need a dry shelter; they hate the rain and cannot tolerate drafts or dampness. A wet goat can catch pneumonia and die. Your goat shed can be a simple, three-sided lean-to or could be converted from an old shed or garage. Make the shed roof watertight and position it so the opening faces away from prevailing winds.

You will need to provide about 15 square feet of space for each goat. A 6 x 8 foot shed provides enough room for two does and their offspring. Provide plenty of clean bedding, such as straw, woodchips, or sawdust to absorb urine. The floor should be dirt or concrete—if your goat shed has a wooden floor, use lots of bedding to absorb urine and change it frequently. Each goat will produce about 1 pound of manure for your garden each day.

Your goats will also need an outdoor space or yard. Don't tether your goats, as they can easily get tangled and be injured or strangled. Provide at least 200 square feet of outdoor space for each adult. If you are raising pygmy breeds, they will need 10 square feet of indoor space and 130 square

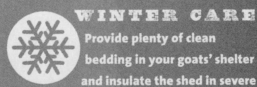

WINTER CARE
Provide plenty of clean bedding in your goats' shelter and insulate the shed in severe climates. Your goats' shelter should be snug and warm; goats are not as hardy as sheep. Close any windows and plug holes so there are no drafts. Make sure water is always available.

Typical Goat Shed

BREEDS
FOR CITY FARMS

There are a number of popular dairy goat breeds that are suitable for city farms. Miniature breeds include **Alpines**, **Oberhaslis**, **Saanens**, **Toggenburgs**, **LaManchas**, and **Nubians**. Adult does of these breeds weigh about 100 to 120 pounds at maturity. **Nigerian Dwarf** dairy goats are considerably smaller, weighing only 30 to 50 pounds at full size. You will need to have at least two goats—they are herd animals and are much happier and easier to manage if they have another goat as a friend. Dogs and cats aren't suitable companions, though a horse may suffice. You will most likely purchase young does that will be bred. Doelings can run $150 or more for purebreds.

Look for healthy animals from a reputable breeder. Ask about the history of related goats, including their temperament, milk production, and ease of kidding. Legs should be straight and the goat should move easily. Check their feet to ensure that they are free of hoof rot. Eyes should be bright, skin soft, and coat smooth. The udder should be well formed, soft, and round, with no scars or injuries. Teats should be of equal size. Make sure their teeth are healthy and the jaw is properly aligned. If your goat has trouble eating, she will not be able to produce as much milk.

feet outside for each goat. Goats are expert escape artists and are hard to contain, so you'll need a sturdy 5-foot-tall fence enclosing your goat yard. Any plants or trees in this area will be eaten. Goats need a shady area and like to climb on things—old doghouses or play structures are great for your goat's play-yard.

RAISING BABIES

Goats are bred once a year and typically produce twins. Does can be bred when they are 10 months old and babies gestate for 5 months. Goats self-deliver, though it is always good to be on hand during the delivery in case something goes wrong. You may choose to disbud your kid's horns, which will stop the horns from growing and potentially harming you or other goats. You'll need to find an experienced goat keeper to help with this. Any male kids will need to be castrated at 6 weeks. Kids are weaned and ready to go to a new home when they are 8 weeks old.

You can bottle-feed baby goats if you want to keep your doe's milk production high. Likewise, you can allow the kids to nurse during the day but separate them from their mothers at night, then milk the doe just once in the morning before you reunite her with the kids.

Angora and cashmere goats are raised for their high-quality fiber. Goats that are grown for fiber look like shaggy dogs with horns. They are typically shorn twice a year; in the spring and fall.

COMMON PROBLEMS

Dairy goats need to be bred to produce milk. Really think out your breeding program. You'll need to locate a suitable buck and make sure you have a plan for the babies. Baby goats are part and parcel of dairy operations. To avoid illnesses and disease, keep housing clean and animals well-fed. Goats may suffer from parasites and intestinal worms—consult your veterinarian about a worming program. Lactating goats may develop mastitis; monitor your lactating does to make sure that their teats aren't infected. If you observe anything unusual, consult your vet.

HARVESTING MILK AND FIBER

Dairy goats need to be milked twice daily. They produce about 900 quarts of milk for 10 months of the year, or roughly 2–6 quarts a day. You'll need a place that is fairly clean and out of the weather for milking. Odors can affect the taste of your milk, so you may want to set up a designated milking room. The ideal milking room is free of dust, dirt, flies, and cobwebs and has a floor that can be kept clean. A milking stand makes it easier to milk. These can be constructed out of scrap wood or purchased from farming supply or feed stores. Give your doe her grain ration while you are milking, then work quickly to finish before she gets bored and wants off the stand.

Goat Terms

Goats are ruminants and have unique names, depending on the age and gender of the animal.

Kid
Baby goat of either gender

Wether
Castrated male

Doeling
Young female

Doe
Adult female

Buck
Adult male

BEES

eeping bees for honey and wax is another way to expand your city farm production. Setting up a hive is easy and takes very little time. Still, there is a lot to know about beekeeping. Weekly maintenance and monitoring are essential to keeping bees healthy, so pick up a good manual, take a class, or find other beekeepers in your area to learn how to manage your hives.

BENEFITS

Honeybees pollinate 90 percent of all fruit and vegetable crops. Keeping a hive increases the pollinators on your farm and may result in more fruit and bigger yields from your garden. You'll be able to harvest honey and beeswax; the surplus can be used as gifts or shared with friends.

BASIC CARE

There are two methods for keeping bees: Langstroth and the Top Bar hives.

Langstroth hives are the most common method of beekeeping. These hives are the typical white boxes that you see on the ends of fields. The hive is like a hanging file cabinet. Frames hang across the box. Each frame has a flat membrane where the bees begin to build their waxy combs. Langstroth hives require special equipment that is difficult to construct by hand, so most beekeepers purchase premade hive components. It could cost between $200 and $400 to set up a couple of Langstroth hives. Purveyors of beekeeping supplies sell beginners' Langstroth hive kits.

Top Bar hives are more accessible to beginning beekeepers and city farmers. They can be constructed out of salvaged materials or old wooden boxes and require only basic building skills. Top Bar hives are low tech and are commonly used in Kenya and Tanzania. For those who want to explore beekeeping, starting a Top Bar hive is easy and inexpensive. This method of beekeeping is gaining in popularity and hive supplies can be purchased online. Many beekeepers use both methods.

Regardless of which method you choose, you will need protective clothing, including gloves, a hat, a veil, and light-colored coveralls. Additionally you'll need a smoker, a bee brush, and a hive tool—this basic gear will run about $150. Package bees (one queen and about 300 workers) to start your hive can be mail-ordered for about $50 or you may be able to get a starter batch from a local beekeeper. Observe bees to make sure they are healthy—they should move and fly easily and that they are not infested with mites (see sidebar).

Put the hive in a quiet, sunny, out-of-the-way place. The area in front of the hive entrance should be open and clear. Bees have a regular flight path to get into their hive, so make sure the opening faces away from roads, paths, or sidewalks. Because your hive will need to be checked weekly throughout the spring and summer, place it so you can easily work all the way around the box.

VARROA DESTRUCTOR MITE

Honeybees have had it rough the past couple of decades. Keepers are observing the mysterious collapse of their colonies. The Varroa mite is certainly giving honeybees a run for their money. This destructive mite is a parasite that feeds on honey, brood, and adults. They are small, reddish, oval discs that are visible on the body of adult bees. With a magnifying glass, they look like a crab.

Healthy hives are infested with Varroa mites by bee drift, hive robbing, and infected package bees. If you keep bees, they will likely have mites, so you should know what to look for and what to do. Novice beekeepers need to do their part in preventing the spread of Varroa mites. Inspect your hive weekly during nectar flow, monitor for mite populations, and use direct controls to reduce the population.

Initially, miticides were used to kill the Varroa, but these are having diminishing effects on new generations that have grown resistant to chemical controls. Beekeepers are developing new approaches for monitoring and reducing mite infestations. As you start beekeeping, learn the life cycle of this dastardly pest, so that you can control it organically.

CONSTRUCTING A TOP BAR HIVE

A Top Bar hive is started in a sturdy box, which can have straight or sloped sides. There are two hive designs: the Kenyan, which has sloped sides, and the Tanzanian, which has straight sides. Both designs work equally well. A straight-sided box is easier to construct and any sturdy wooden box can be converted into a Top Bar hive. Drill 6–10 holes ½-inch in diameter on one end or on the side of your hive box, so bees can enter and exit the hive. Mount your hive on legs so it is raised 3–4 feet above the ground. Make sure the hive box is watertight and painted with three coats of exterior paint.

After you have your hive box, top bars must be cut and prepared. Top bars run across the hive box and the comb hangs from each bar. The width and spacing of the top bars is important. Top bars should be 1¼–1½ inches wide. This allows the proper space between the combs for the bees.

Top bars are prepared so that bees can use them to build their comb. Each bar has a groove that runs down the middle. A piece of hardwood, masonite, or beeswax is wedged into this slot. The bees start to build their comb on this strip. Leave a 2-inch space between the side of the hive and the end of the wood or beeswax strip—this will keep bees from attaching their comb to the sides of your hive box. Top bars can be prepared with a thick rope of beeswax wedged into the groove or with a bead of melted wax down the middle of the bar.

Most keepers put some kind of roof on their hives. Curved metal roofs, painted white to reflect heat, or corrugated plastic roofing can be put on top of the hive and held down with bricks. If you are using metal, make sure that it sits on a frame above the hive so that the heat conducted by the metal doesn't overheat your bees.

Introduce or attract bees to your hive and start beekeeping. Package bees can be purchased to start your colony—instructions for introducing bees come with your package. You may be able to attract bees to your hive by painting the inside of the hive box with propolis and ammonia. Propolis is the sticky substance that bees collect from the surface of leaves—after the ammonia is dried, the smell of propolis is irresistible to migratory bees. You may be able to get propolis from a local beekeeper or from a purveyor of apiary supplies.

Your hive will need to be opened and monitored each week as the season progresses and bees start to build their comb, Don your protective clothing, fire up the smoker, and see how the colony is progressing. When you work around bees, move slowly and calm your mind. Use a little smoke to subdue the bees. Lift off the roof. Using a hive tool, pry loose the top bars at the end of the hive opposite the entrance holes. Carefully lift up the top bar and keep it horizontal so that the fragile comb doesn't break off.

Top bar honey is whole-comb harvested, which means you won't be extracting the honey as you would with the Langstroth hives. You will drain or press the honey from the comb, a process that imparts more flavor to the

Typical Top Bar Hive

honey. Harvest the wax by melting and straining it through a paper towel, letting it drain onto water.

You will get stung when you are working with bees. Most beekeepers just shrug this off as something that goes with the territory. If you are fearful of being stung or are allergic to bee stings, then beekeeping may not be for you. If you are stung, don't pinch the stinger to pull it out—this only pushes more venom into your skin. Instead, use a credit card or fingernail to scrape the stinger off.

BASIC CARE

Most adult honeybees live just six weeks during the peak of nectar production in your garden.

Workers that pupate near the end of summer overwinter in the hive, while most of the other worker bees and drones die in the field. Queens will typically live two to three years.

The hive is active and bees are gathering food from spring though summer, with a peak nectar flow near the end of the growing season. An industrious colony will fill all the space in the hive box. When this happens, additional frames or top bars are stacked on top. These supers expand space to store food and increase honey production.

Bees need enough food to last the winter, so make sure you leave half of the nectar bars for the colony that will overwinter. Entrance holes can be closed to keep bees warmer, and the hive can be insulated for winter. Check with your local beekeepers' association for the recommended winterizing techniques for your area.

HARVESTING HONEY

You will find combs that are filled with honey and capped with beeswax. Harvest the comb when most of the cells are filled and capped. You will need some sort of frame so that the top bar and comb can hang outside the hive and you can inspect it. This can be easily constructed with heavy wire mounted in a wide, thick board.

COMMON PROBLEMS

Bears, raccoons, ants, and other insects may want to share in your honey yield. Put the legs of your hive in cans of water to keep insects from crawling into the hives. For larger animals, consult your local beekeepers' association for local pest control strategies.

Animals on the Farm

Raising livestock in the city is a great family activity and a way to increase the kinds of food you can produce just beyond your doorstep. Be clear about your goals for keeping livestock and make sure you have adequate space and time to provide optimal care for your animals. Check your municipal land use codes to make sure that keeping small farm animals is allowed in your city.

Profile:
JUSTICE FOR GOATS

It's a short trip from chickens in your backyard to goats in the shed—just ask Jennie Grant, a Seattle gardener who raises chickens, keeps bees, and also has goats. Jennie established the Goat Justice League to educate city gardeners (and city officials) about the possibilities of urban goat tending.

The Goat Justice League provides information for those curious about what it takes to raise goats in the city. Not only are they social animals, but they provide milk, which can be turned into cheese and yogurt. On her Web site, Jennie goes through goat keeping step-by-step, giving the basics, plus telling tales of her city farm life. Because she names all her animals (even the bees), it's easy to feel as if you are a friend of the farm.

Of course, not every city thinks you should keep goats—some don't even let you keep chickens. The Goat Justice League campaigned to give Seattle citizens the chance to keep goats, and it worked. Homeowners can now keep up to three miniature goats.

How big is a miniature goat? Mini goats are a cross between dwarf Nigerian and another breed. They are good for milk, and a great choice for city farms, because they top out at about 100–120 pounds. Now to figure out where you will put their shed.

LEARN MORE: goatjusticeleague.org

RESOURCES

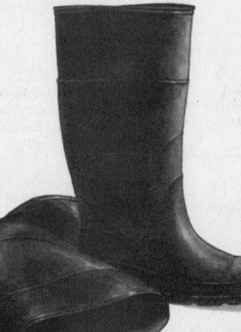

 GATHER YOUR TOOLS and assemble your resources. Seek out resources in your community that can help you grow successfully. Talk to friends and neighbors who garden and find out what they grow and what does well in your climate. Check out the local planting calendar (available through a local garden club or your cooperative extension service). The Internet is an excellent and ever-expanding source of information about city farming. Collect other resources, such as seed suppliers, compost sources, tool suppliers, your local garden hotline, and organizations that give classes on urban farming. Adapt what you learn to your site, and try it! Here are some favorite resources, sources for supplies and seeds, information about tools, and an glossary of organic gardening terms.

Books

General topics

Colebrook, Binda. **Winter Gardening in the Maritime Northwest: Cool-Season Crops for the Year-Round Gardener.** Seattle: Sasquatch Books, 1998.

Coleman, Eliot. **Four Season Harvest: Organic Vegetables from Your Home Garden All Year Long.** Illustrations by Kathy Bray; photographs by Barbara Damrosch. White River Junction, Vt.: Chelsea Green Publishing, 1999.

Creasy, Rosalind. **The Complete Book of Edible Landscaping.** Illustrated by Marcia Kier-Hawthorne. San Francisco: Sierra Club Books, 1982.

Larkcom, Joy. **Creative Vegetable Gardening**. London: Mitchell Beaszley, 2008.

The Maritime Northwest Garden Guide. Seattle: Seattle Tilth, 2008.

Otto, Stella. **The Backyard Orchardist: A Complete Guide to Growing Fruit Trees in the Home Garden**. Maple City, Mich.: Otto Graphics, 1993.

Proulx, E. Annie. **The Fine Art of Salad Gardening.** Emmaus, Pa.: Rodale Press, 1985.

Seymour, John. **The Self-Sufficient Gardener: A Complete Guide to Growing and Preserving All Your Food**. Garden City, N.Y.: Dolphin Books, 1978.

Solomon, Steve. **Growing Vegetables West of the Cascades: The Complete Guide to Organic Gardening.** Seattle: Sasquatch Books, 2007.

Soil and composting

Gershuny, Grace. **Start with the Soil: The Organic Gardener's Guide to Improving Soil for Higher Yields, More Beautiful Flowers, and a Healthy, Easy-Care Garden**. Emmaus, Pa.: Rodale Press, 1993.

Martin, Deborah L. & Grace Gershuny, eds. **The Rodale Book of Composting.** Emmaus, Pa.: Rodale Press, 1992.

Stell, Elizabeth. **Secrets to Great Soil: A Grower's Guide to Composting, Mulching and Creating Healthy, Fertile Soil for Your Garden and Lawn.** Pownal, Vt.: Storey Communications, Inc., 1998.

Seeds

Coulter, Lynn. **Gardening with Heirloom Seeds: Tried-and-True Flowers, Fruits and Vegetables for a New Generation.** Durham: University of North Carolina Press, 2006.

McVicar, Jekka. **Seeds: The Ultimate Guide to Growing Successfully from Seed.** London: Kyle Cathie, 2008.

Rogers, Marc. **Saving Seeds: The Gardener's Guide to Growing and Storing Vegetable and Flower Seeds**. Illustrations by Polly Alexander. North Adams, Ma.: Storey Publishing, 1990.

Weeds and wild plants

Elias, Thomas S. & Peter A. Dykeman. **Edible Wild Plants: A North American Field Guide.** New York: Sterling Publishing Co., 1990.

Jacobson, Arthur Lee. **Wild Plants of Greater Seattle**. Seattle: Arthur Lee Jacobson, 2001.

Western Society of Weed Science. **Weeds of the West.** Jackson: University of Wyoming, 1996.

Maintenance

Brickell, Christopher & David Joyce. **The American Horticultural Society Pruning and Training.** New York: DKPublishing Inc., 1996.

Kourik, Robert. **Drip Irrigation for Every Landscape and All Climates: Helping Your Garden Flourish While Conserving Water!: Outwit Droughts with Expert Guidance**. White River Junction, Vt.: Metamorphic Press (Chelsea Green), 2009.

Animals

Chandler, P. J. **The Barefoot Beekeeper: Low Cost, Low Impact Natural Beekeeping for Everyone**. www.biobees.com: Philip Chandler, 2009.

Damerow, Gail, ed. **Barnyard in Your Backyard: A Beginner's Guide to Raising Chickens, Ducks, Geese, Rabbits, Goats, Sheep, and Cattle**. North Adams, Ma.: Storey Publishing, 2002.

Thomas, Steven & George P. Looby, DVM. **Backyard Livestock: Raising Good Natural Food for Your Family**, 3rd ed. Illustrations by Mark Howell & Patricia Witten. Woodstock, Vt.: The Countryman Press, 2007.

Insects, pests, and diseases

Carr, Anna. **Rodale's Color Handbook of Garden Insects.** Emmaus, Pa.: Rodale Press, 1979.

Dreistadt, Steve H. **Pests of Landscape Trees and Shrubs: An Integrated Pest Management Guide.** University of California Division of Agriculture and Natural Resources, Publication 3359, 1994.

Flint, Mary Louise. **Pests of the Garden and Small Farm: A Grower's Guide to Using Less Pesticides.** University of California Division of Agriculture and Natural Resources, Publication 3332, 1990.

Olkowski, William; Sheila Daar; and Helga Olkowski. **Common-Sense Pest Control.** Newtown, Conn.: The Taunton Press, 1991.

Using and preserving produce

Ball Corporation. **Ball Blue Book: The Guide to Home Canning and Freezing**. Muncie, Ind.: Ball Corporation, 1983.

McClure, Susan, and the Rodale Food Center. **Preserving Summer's Bounty: A Quick and Easy Guide to Canning, Preserving and Drying What You Grow.** Emmaus, Pa.: Rodale Press, 1998.

Morash, Marian. **The Victory Garden Cookbook.** New York: Alfred A. Knopf, 1982.

Waters, Alice. **Chez Panisse Vegetables**. New York: HarperCollins Publishers, 1996.

Alternative Agriculture Resources

Biointensive gardening

Jeavons, John. **How to Grow More Vegetables (and Fruits, Nuts, Berries, Grains, and Other Crops) Than You Ever Thought Possible on Less Land Than You Can Imagine.** Berkeley, Calif.: Ten Speed Press, 2006.

Martin, Orin. **French Intensive Gardening: A Retrospective. News and Notes of the UCSC Farm and Garden,** Issue 112 (Winter 2007).

www.growbiointensive.org

No till or natural farming

Fukuoka, Masanobu. **The Natural Way of Farming: The Theory and Practice of Green Philosophy**. Translated by Frederic P. Metreaud. Tokyo: Japan Publications, Inc., 1985.

Fukuoka, Masanobu. **The One-Straw Revolution: An Introduction to Natural Farming**. New York: New York Review Books, 2009.

Stout, Ruth. **Gardening without Work: For the Aging, the Busy and the Indolent.** Line drawings by Nan Stout. New York: Lyons Press, 1998.

http://www.notill.org/

Biodynamic farming

Biodynamic Farming and Gardening Association **www.biodynamics.com**

Demeter-International e. V., International Biodynamic certification organization **www.demeter.net**

Diver, Steve. **Biodynamic Farming and Compost Preparation.** Appropriate Technology Transfer for Rural Areas **http://attra.ncat.org/attra-pub/biodynamic. html#appendix3**

Josephine Porter Institute. **Working with the Stars: A Bio-Dynamic Sowing and Planting Calendar.** Woolvine, Va.: Josephine Porter Institute for Applied Biodynamics, Inc., published yearly. **www.jpibiodynamics.org**

Stella Natura: Biodynamic Planting Calendar. Sherry Wildfeuer, ed. Kimberton, Pa.: Camphill Village Kimberton Hills, published each year. **www.stellanatura.com**

Permaculture

Hemenway, Toby. **Gaia's Garden: A Guide to Home-Scale Permaculture.** White River Junction, Vt.: Chelsea Green Publishing Company, 2001.

Permaculture Institute **www.permaculture.org/**

Permaculture Activist **www.permacultureactivist.net/**

Sources for Farm Supplies

A. M. Leonard **www.amleo.com**

Drip Works **www.dripworksusa.com**

Lee Valley Tools **www.leevalley.com**

Peaceful Valley Farm Supply **www.groworganic.com**

Red Pig Tools **www.redpigtools.com**

Terrebonne Limited **www.terrebonnelimited.com**

Yelm Earthworm Farm **www.yelmworms.com**

Seed and Plant Sources

Abundant Life Seeds **www.abundantlifeseeds.com**

Baker Creek Heirloom Seeds **http://rareseeds.com**

Bountiful Gardens **www.bountifulgardens.org**

Fedco Seeds **www.fedcoseeds.com/seeds.htm**

Heirloom Seeds **www.heirloomseeds.com**

Native Seeds/S.E.A.R.C.H. **www.nativeseeds.org**

One Green World **www.onegreenworld.com**

Peter's Seed and Research **www.psrseed.com**

Raintree Nursery
www.raintreenursery.com

Renee's Garden Seeds
www.reneesgarden.com

Seed Savers Exchange
www.seedsavers.org

Seeds Trust **www.seedstrust.com**

Southern Exposure Seed Exchange
www.southernexposure.com

Organizations

American Community Gardening Association
www.communitygarden.org

ATTRA: Appropriate Technology Transfer for Rural Areas **www.attra.org**

City Farmer, Canada's Office of Urban Agriculture **www.cityfarmer.org**

Gardening Organic (previously known as the Henry Doubleday Research Association)
www.gardenorganic.org.uk

Kitchen Gardeners **www.kitchengardeners.com**

Organic Agriculture (a part of the Food and Agriculture Organization of the United Nations)
www.fao.org/organicag/en

Seattle Tilth **www.seattletilth.org**

STATE ORGANIC FARMERS AND GROWERS ASSOCIATIONS

Maine Organic Farmers and Gardeners Association
www.mofga.org

Florida Organic Growers **www.foginfo.org**

California Certified Organic Farmers **www.ccof.org**

SafeLawns.org

Seeds of Diversity (Canada's seed saving network) **www.seeds.ca/en.php**

Thrive (Britain's organization to promote gardening accessibility)
www.thrive.org.uk

Tools

Which tools should you buy? You don't need every gizmo under the sun—the newest gadget won't make your garden grow any better. A solid collection of a few high-quality tools will take you far. Here are the tools to buy:

Digging fork:
The tines on this tool are square or triangular, which helps it get into soil and break it up.

Digging spade:
A digging or border spade has a flat or only slightly rounded tip. Spades are used for digging a garden bed and creating straight edges.

American shovel:
Typical, usually long-handled, all-purpose shovel with a pointed tip. Shovels are good at digging holes and piling stuff up.

Hori hori:
A Japanese-style knife that makes a good all-purpose hand tool for weeding or creating furrows for planting

Hand trowel:
It's like a little shovel in your hand.

Steel rake:
Use this for smoothing out the soil before you plant seeds and for raking out rocks from the garden bed. Steel rakes have short, curved tines attached to a horizontal metal piece.

Leaf rake:
Triangular in shape and made from aluminum, bamboo, or some other bendable material, leaf rakes do a good job of, well, raking leaves.

Pitchfork:
Need a pile of hay moved? Are you turning the compost pile? A pitchfork has widely spaced, slightly curved, slender round tines with pointed tips. Don't try to use a pitchfork to dig in the soil—or vice versa.

Hand pruners:
Choose bypass pruners, which cut the stem when the sharp blade passes by the anchor blade. Anvil pruners cut by pushing the sharp blade against the anvil, smashing the stem. Hand pruners come in several different sizes, so be sure to try them out at a nursery before purchasing.

Lopping shears:
Bypass lopping shears do the same thing as hand pruners, only on bigger branches and from farther away. The usual rule of thumb for hand pruners is that you should cut nothing larger than the thickness of your thumb. Loppers will cut through thicker branches, and their long handles help you prune higher up in the apple tree.

Scissors:

Scissors are a wonderful tool in the garden, snipping everything from lettuce leaves to twine. Be sure to buy a pair with short blades, so you cut only the leaves you want. Stainless steel blades won't rust.

Scuffle hoe:

You can find scuffle hoes in different shapes; some have a circle or stirrup of metal, or are shaped like two wings attached at a point. A scuffle hoe cuts off annual weeds at or just under the soil surface when you "scuffle" it back and forth. Good for paths or around shallow-rooted plants.

Other handy stuff

Scoop shovel:

It's almost like a snow shovel, and will help you pick up large amounts of loose material, such as woodchips or potting soil.

Harvest basket or trug:

Choose from metal or a long-lasting wood, such as cedar. You can rinse produce right in the metal harvest basket in an outdoor sink.

Wheelbarrow or garden cart:

If you have enough room in the garden, it's handy to be able to move around large amounts of soil or

mulch. Wheelbarrows have three wheels, and can tip over if the load is not balanced; garden carts with four wheels are more stable.

Soil thermometer:

Is it time to plant those bean seeds? A soil thermometer is a handy tool that helps you plan when to plant.

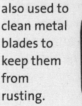

Weed trimmer:

For those areas of leftover lawn where a lawn mower might not fit.

Sandpaper, steel wool, disposable plastic gloves, dust mask:

All supplies that help you keep your garden tools shipshape and in order.

Linseed oil:

Made from flax, can be rubbed into wood handles to help preserve them and also used to clean metal blades to keep them from rusting.

Sharpening stone and file:

To keep the edges of your pruners and shovels in good condition.

Tool-buying tips

- Look for tools that have been forged from one solid piece of metal, from the tip of the tool head to high up around the handle, instead of two pieces of metal welded together (the latter can break easily). Tools forged from one piece of metal are more expensive, but will last for many years.

- The metal that's wrapped around the end of the handle is held together with rivets—two sets of rivets will be stronger.

- Stainless steel digging forks and spades resist rust, and mud doesn't stick to them easily.

- Short or long handle? That's up to you—the ease and comfort may depend on how tall you are. Long-handled tools may give you better leverage, but if you're short, the handle may get in the way.

- D-handles have a grip at the top of the handle, making it easy to grab.

HANDLE

HEAD

How to file a spade or shovel

Make your work easier by keeping your tools sharp. Use a whetstone or sharpening tool for your hand pruners and lopping shears. Use a fine file to remove and smooth the edge of your shovel.

Take care of your tools, and your tools will take care of your garden

How to clean your tools:

- Always knock off, wipe off, or wash off soil that sticks to tools.

- Dry off head and handle before storing.

- Make sure tools are kept dry and not left out in the rain.

- At season's end (or when there's a breathing space in your digging and harvesting), clean off tools, sharpen, oil, and store them.

- Store your shovel, spade, and digging fork for the winter by plunging them in a 5-gallon bucket of sand to which you've added ½ gallon of linseed or mineral oil. Sand wooden handles and apply a thin coat of mineral oil (wipe off any extra).

How to use your tools without breaking your back

Here are some techniques for making tools work for you.

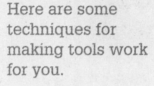

■ *Carry the load across your body. Hold your shovel or pitchfork so that the load moves across your body on the way to the wheelbarrow, rather than out and away from you.*

■ *It's a breeze, if you bend at the knees. Keep your back straight and bend your knees when digging and lifting material.*

■ *Use your legs when working a digging fork or spade into the soil—rather than stabbing the soil, use your arms to steady the tool and your legs to do the work.*

■ *Switch your grip. Save your wrist by turning your hand trowel or hori hori upside down.*

Sample Farmsteads

Containers on a deck or patio

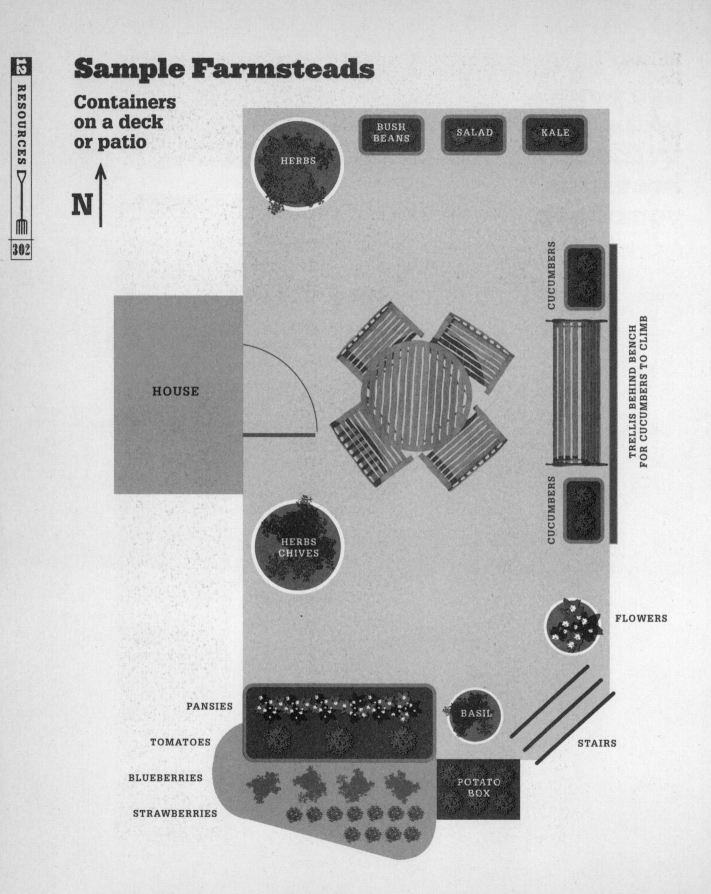

N

HERBS

BUSH BEANS

SALAD

KALE

CUCUMBERS

CUCUMBERS

TRELLIS BEHIND BENCH
FOR CUCUMBERS TO CLIMB

HOUSE

HERBS
CHIVES

FLOWERS

PANSIES

TOMATOES

BLUEBERRIES

STRAWBERRIES

BASIL

STAIRS

POTATO
BOX

Raised Beds

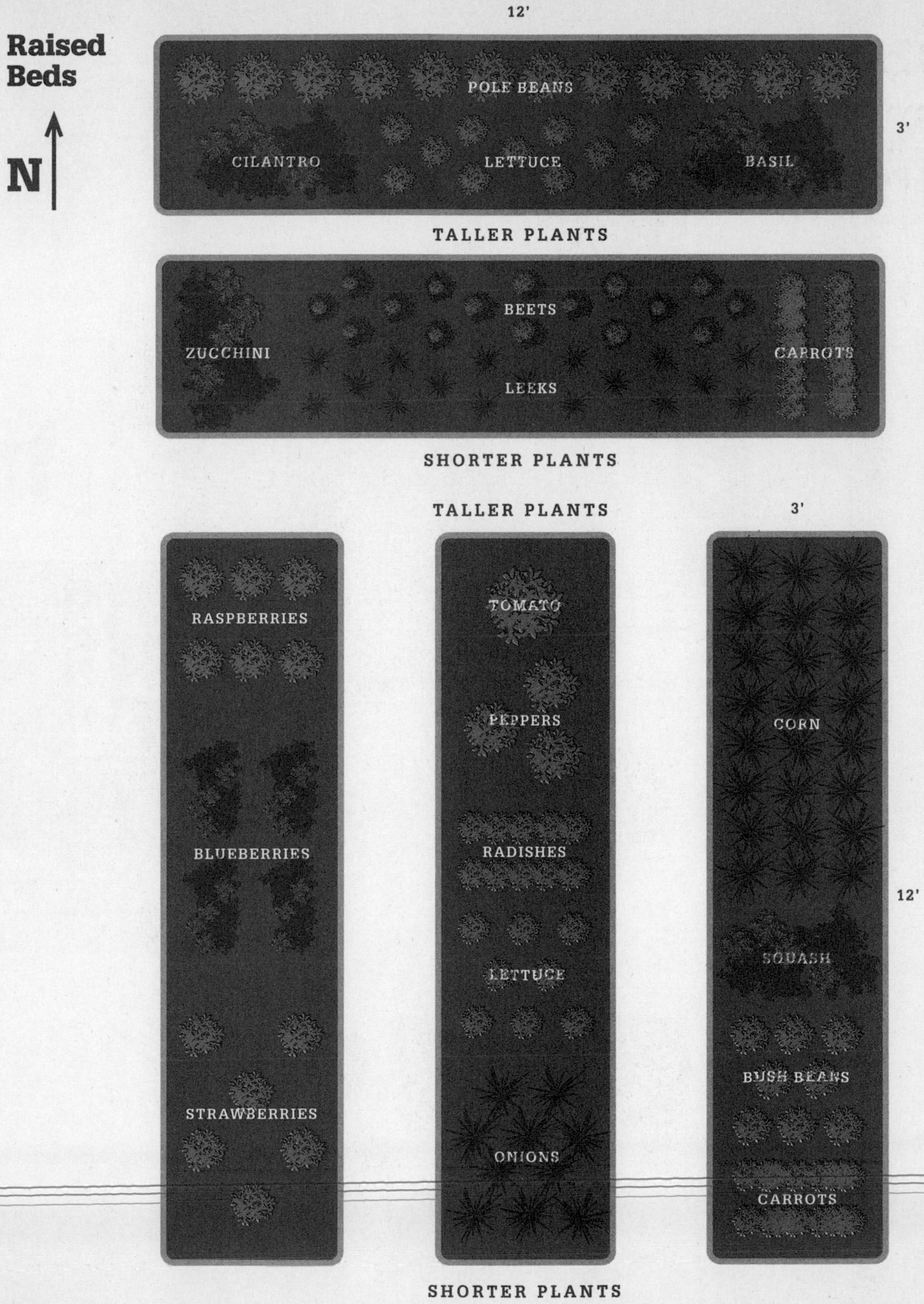

N ↑

12'

3'

POLE BEANS

CILANTRO   LETTUCE   BASIL

TALLER PLANTS

BEETS

ZUCCHINI   LEEKS   CARROTS

SHORTER PLANTS

TALLER PLANTS

3'

RASPBERRIES

BLUEBERRIES

STRAWBERRIES

TOMATO

PEPPERS

RADISHES

LETTUCE

ONIONS

CORN

12'

SQUASH

BUSH BEANS

CARROTS

SHORTER PLANTS

Sample of large suburban lot (1/4 – 2 acres)

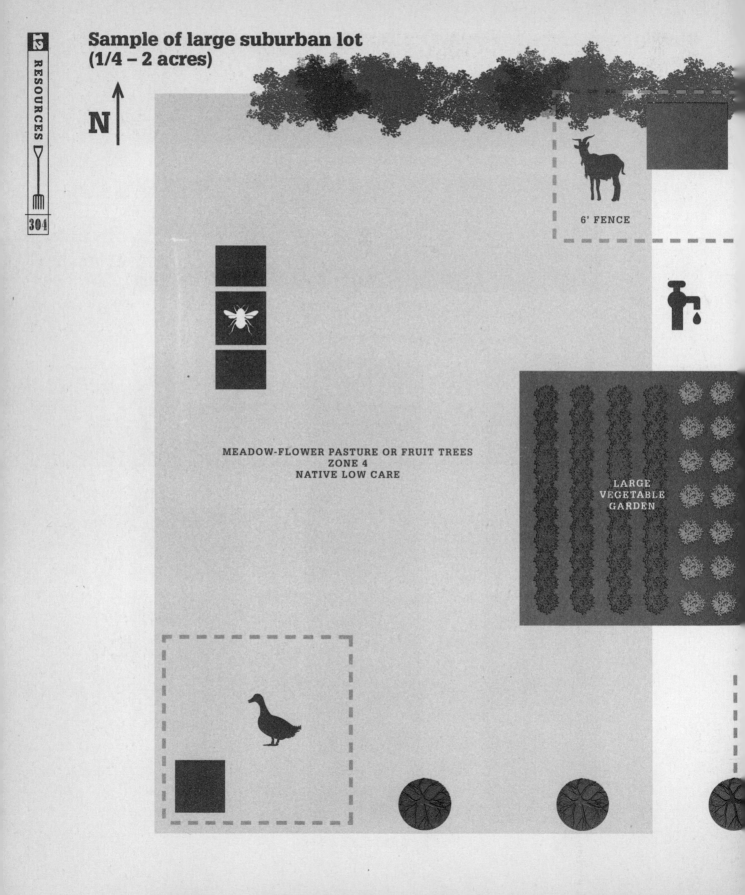

N

6' FENCE

MEADOW-FLOWER PASTURE OR FRUIT TREES
ZONE 4
NATIVE LOW CARE

LARGE
VEGETABLE
GARDEN

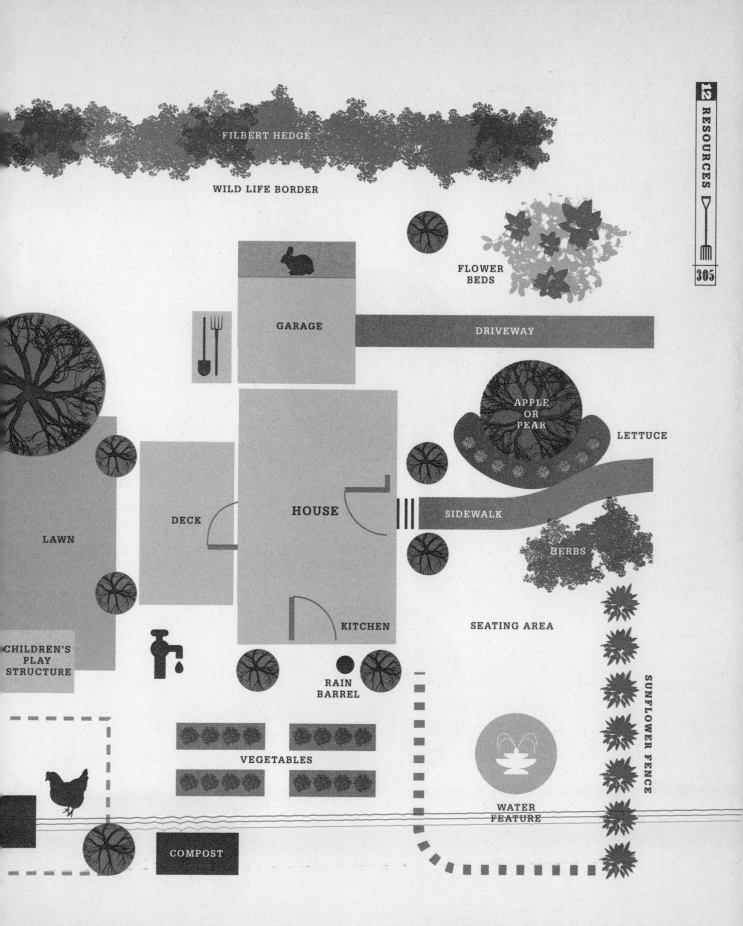

FILBERT HEDGE

WILD LIFE BORDER

FLOWER BEDS

GARAGE

DRIVEWAY

APPLE OR PEAR

LETTUCE

HOUSE

DECK

SIDEWALK

HERBS

LAWN

CHILDREN'S PLAY STRUCTURE

KITCHEN

SEATING AREA

RAIN BARREL

VEGETABLES

WATER FEATURE

SUNFLOWER FENCE

COMPOST

**Typical urban lot
(6500-8500 sq ft)
Starting place**

N

GARAGE

KITCHEN

ARBOR VITAE HEDGE

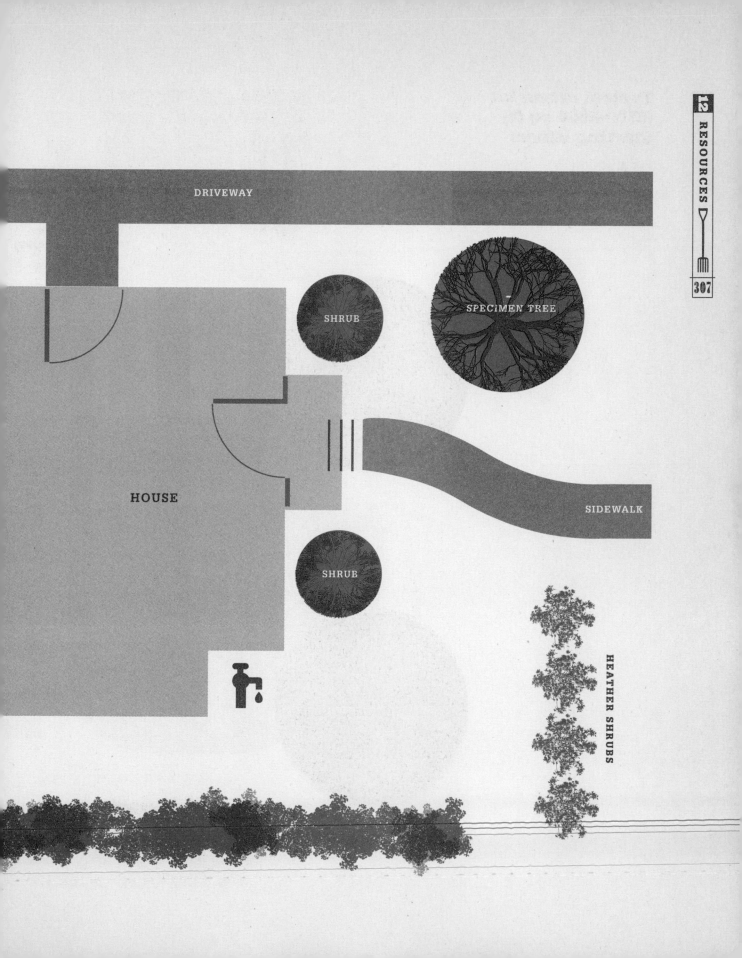

DRIVEWAY

SHRUB

SPECIMEN TREE

HOUSE

SIDEWALK

SHRUB

HEATHER SHRUBS

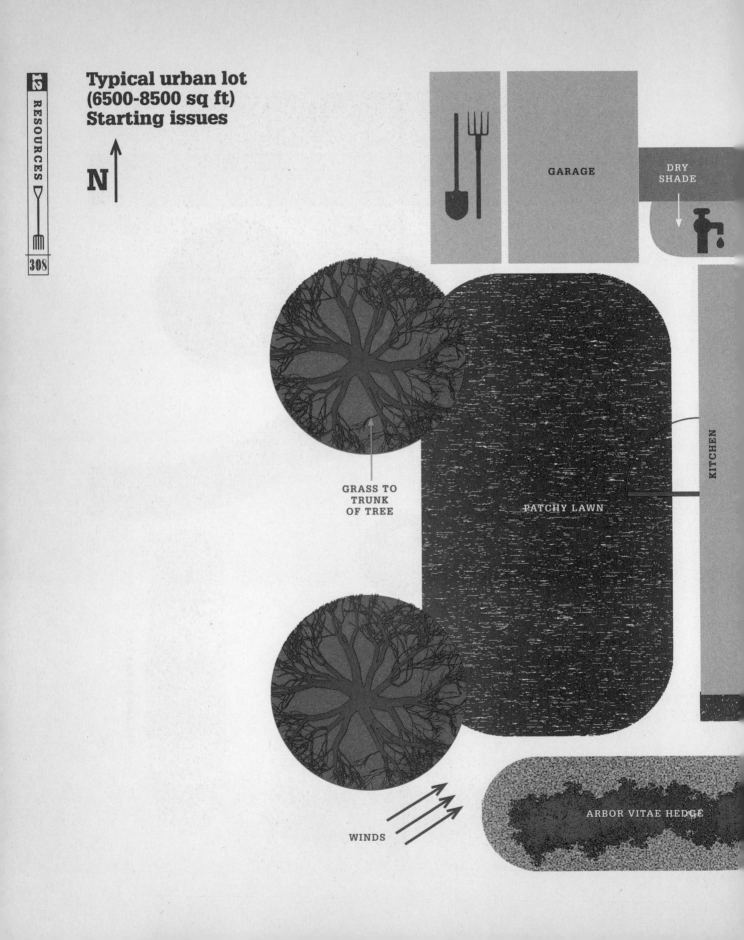

**Typical urban lot
(6500-8500 sq ft)
Starting issues**

N

GARAGE

DRY
SHADE

GRASS TO
TRUNK
OF TREE

PATCHY LAWN

KITCHEN

WINDS

ARBOR VITAE HEDGE

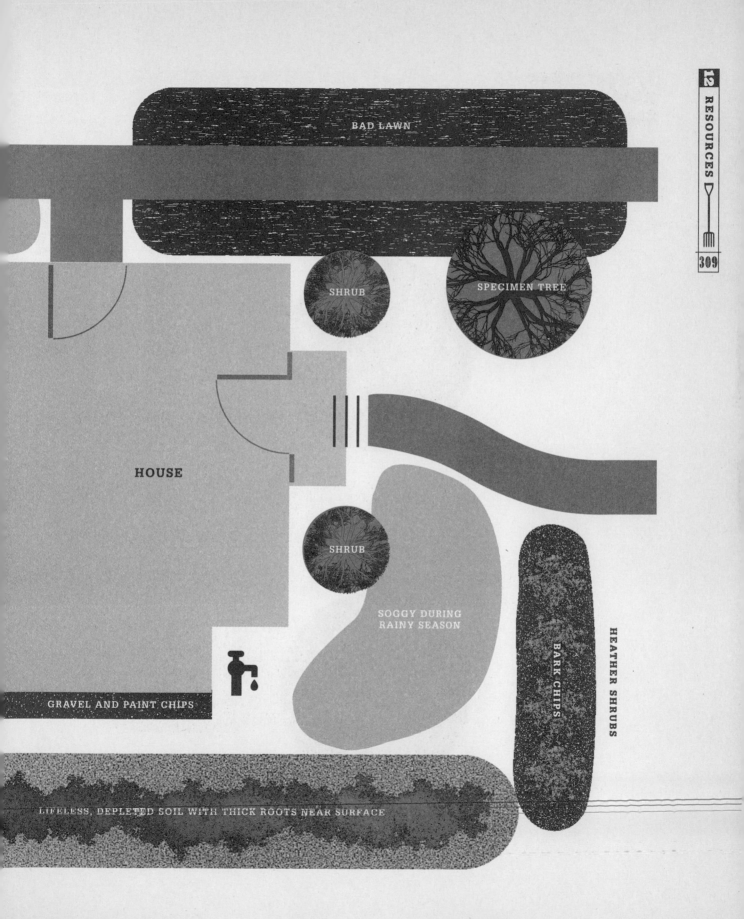

BAD LAWN

SHRUB

SPECIMEN TREE

HOUSE

SHRUB

SOGGY DURING
RAINY SEASON

GRAVEL AND PAINT CHIPS

BARK CHIPS

HEATHER SHRUBS

LIFELESS, DEPLETED SOIL WITH THICK ROOTS NEAR SURFACE

**Typical urban lot
(6500-8500 sq ft)
Zones**

N

ZONE 2

GARAGE

KITCHEN

ZONE 5   ZONE 3   ZONE 2   ZONE 1

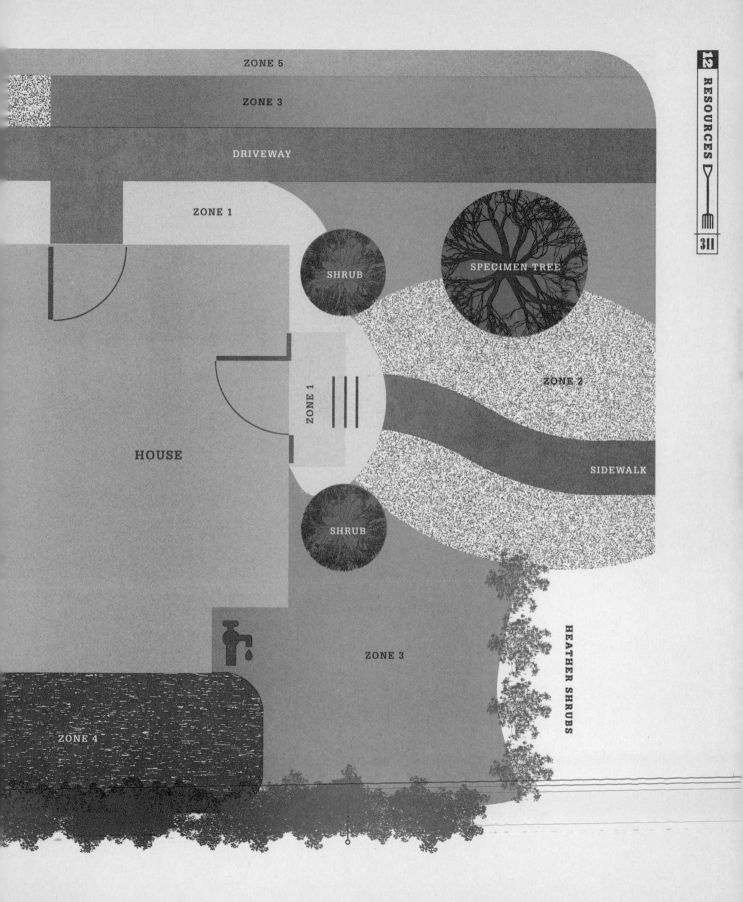

ZONE 5

ZONE 3

DRIVEWAY

ZONE 1

SHRUB

SPECIMEN TREE

ZONE 1

HOUSE

ZONE 2

SIDEWALK

SHRUB

HEATHER SHRUBS

ZONE 3

ZONE 4

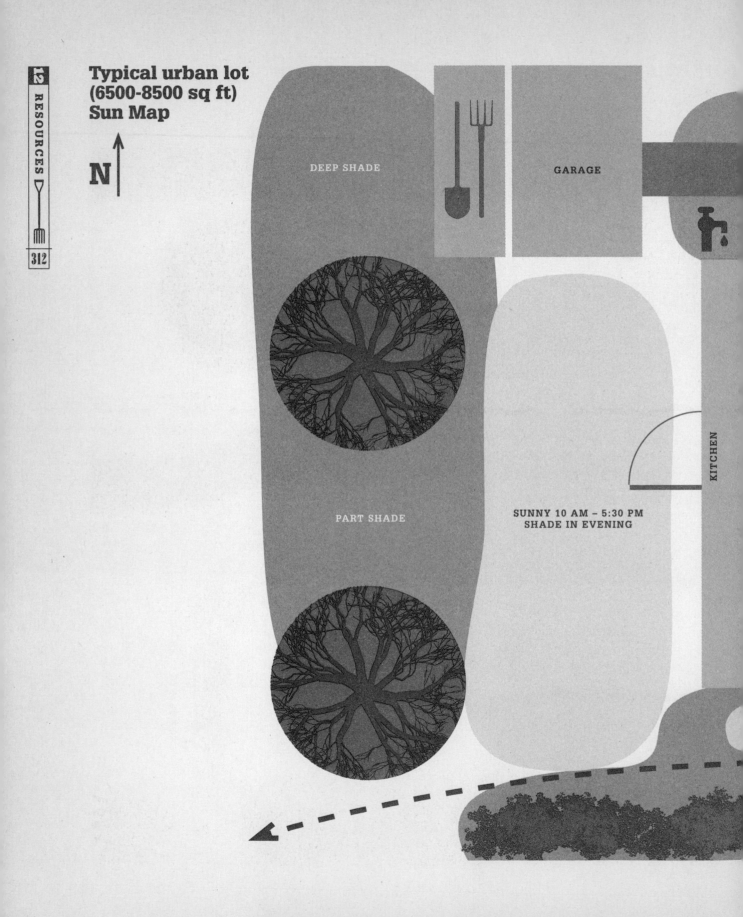

Typical urban lot (6500-8500 sq ft) Sun Map

N

DEEP SHADE

GARAGE

PART SHADE

SUNNY 10 AM – 5:30 PM
SHADE IN EVENING

KITCHEN

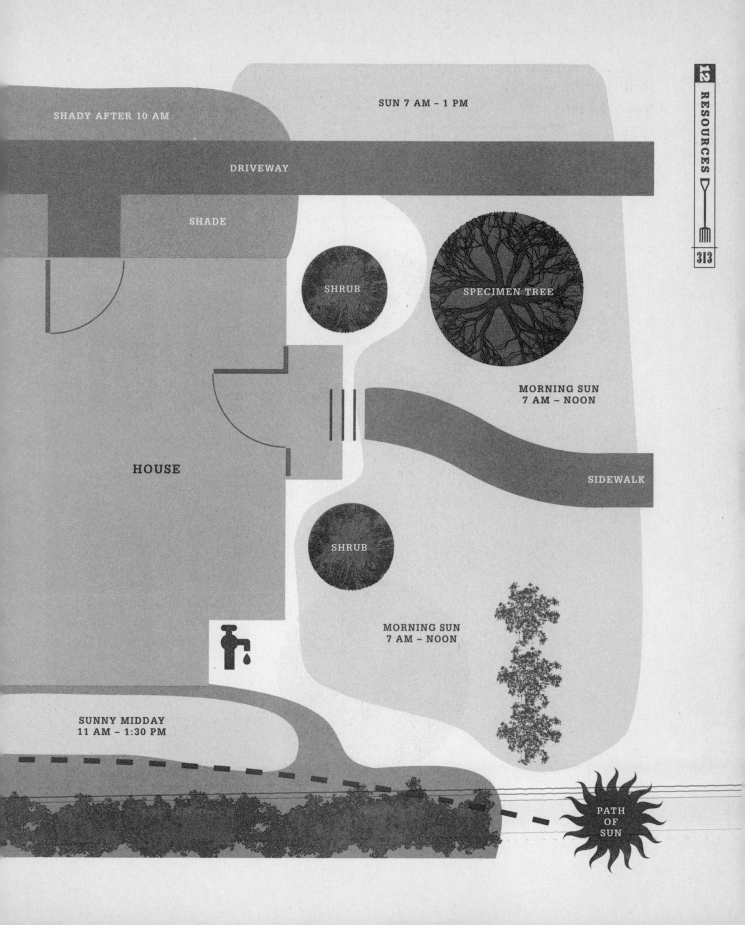

SHADY AFTER 10 AM

SUN 7 AM – 1 PM

DRIVEWAY

SHADE

SHRUB

SPECIMEN TREE

MORNING SUN
7 AM – NOON

HOUSE

SIDEWALK

SHRUB

MORNING SUN
7 AM – NOON

SUNNY MIDDAY
11 AM – 1:30 PM

PATH
OF
SUN

Typical urban lot (6500-8500 sq ft) Year 1

N ↑

Back yard:

- Added two 3' x 8' veggie beds
- Sheet mulched two more future beds
- Started compost bin and worm bin

Front yard:

- Added front herb garden
- Added two large containers for salad
- Added two large containers with edible flowers

GARAGE

WORM BIN

VEGETABLES

KITCHEN

SHEET MULCH

COMPOST

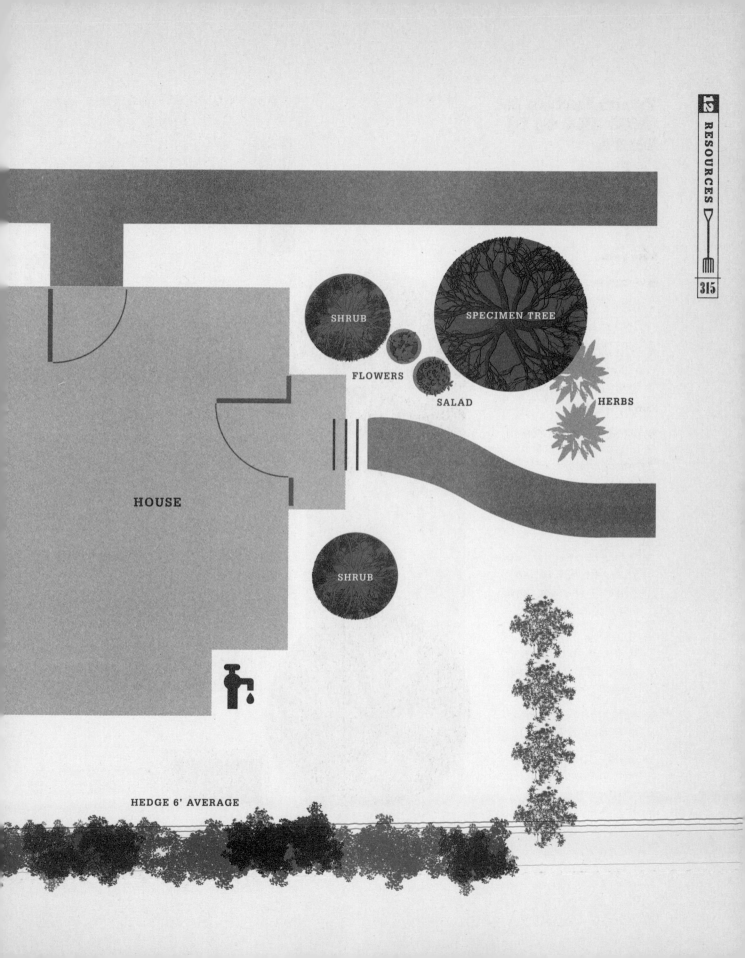

SPECIMEN TREE

SHRUB

FLOWERS

SALAD

HERBS

HOUSE

SHRUB

HEDGE 6' AVERAGE

Typical urban lot (6500-8500 sq ft) 2-3 Years

N ↑

Year two:

- Cultivate two new veggie beds (sheet mulched in year one)

- Mulched around large trees in back yard

- Planted two grapes in huge containers

- Created clover footpath around front and South side of house

Year three:

- Added rain barrels

- Ripped out heather shrubs and planted herbs and flowers—enlarged flower bed by specimen tree.

- Ripped out shrub (South of front door) and created a flower bed for pollinators.

- Added stepping stones in "pollinator" bed

- Mulched front specimen tree and area north of driveway

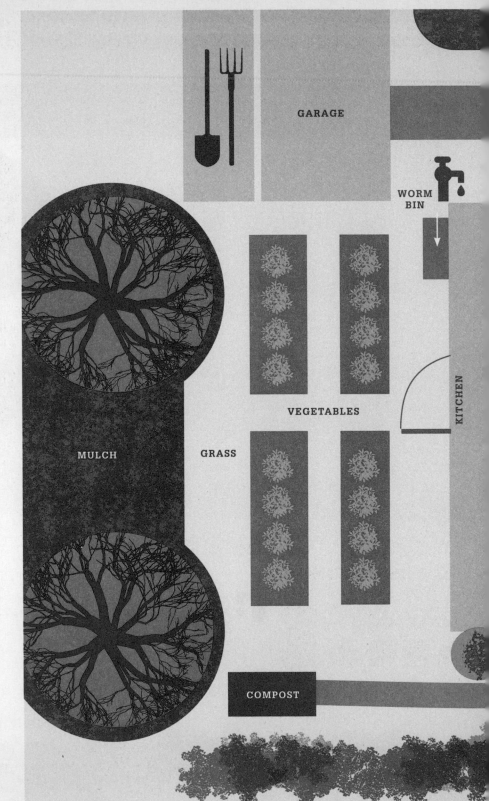

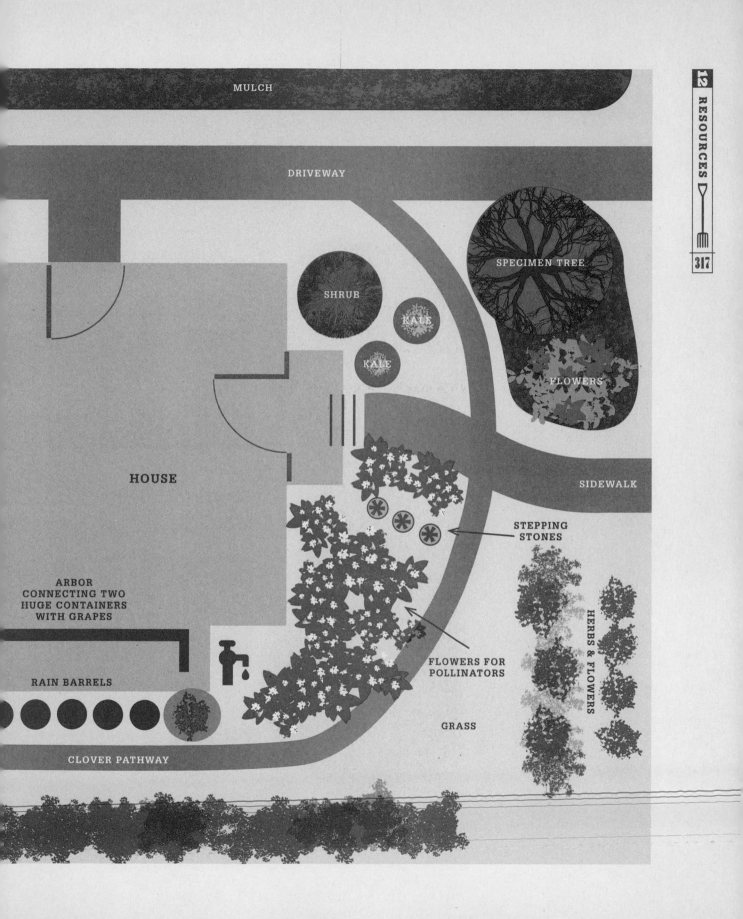

MULCH

DRIVEWAY

SPECIMEN TREE

SHRUB

KALE

KALE

FLOWERS

HOUSE

SIDEWALK

STEPPING
STONES

FLOWERS FOR
POLLINATORS

HERBS & FLOWERS

ARBOR
CONNECTING TWO
HUGE CONTAINERS
WITH GRAPES

RAIN BARRELS

GRASS

CLOVER PATHWAY

Typical urban lot (6500-8500 sq ft) Years 4 plus

N ↑

Year four:

■ *Added water fountain with small pond in front and birdbath*

■ *Ripped out shrub (North of front door) and made large keyhole veggie bed*

■ *Planted apple tree fence or "cordon"*

■ *Added beehives and chicken coop (or rabbit hutch)*

■ *Enlarged herb and flower beds*

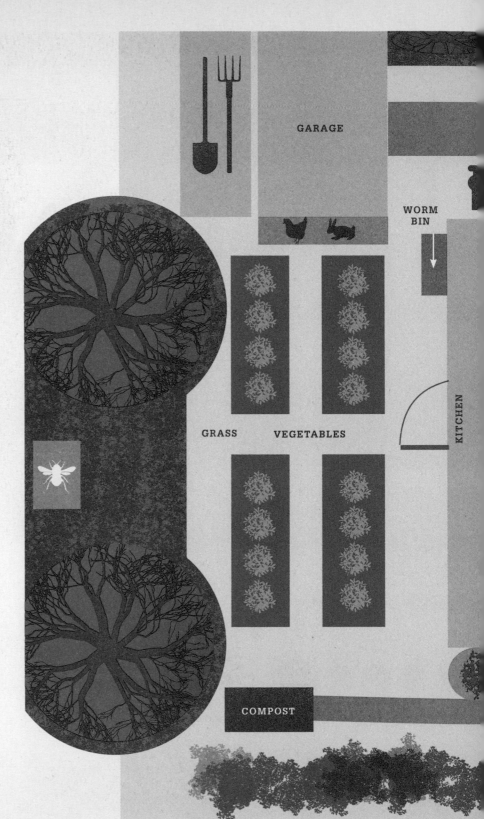

GARAGE

WORM BIN

GRASS   VEGETABLES

KITCHEN

COMPOST

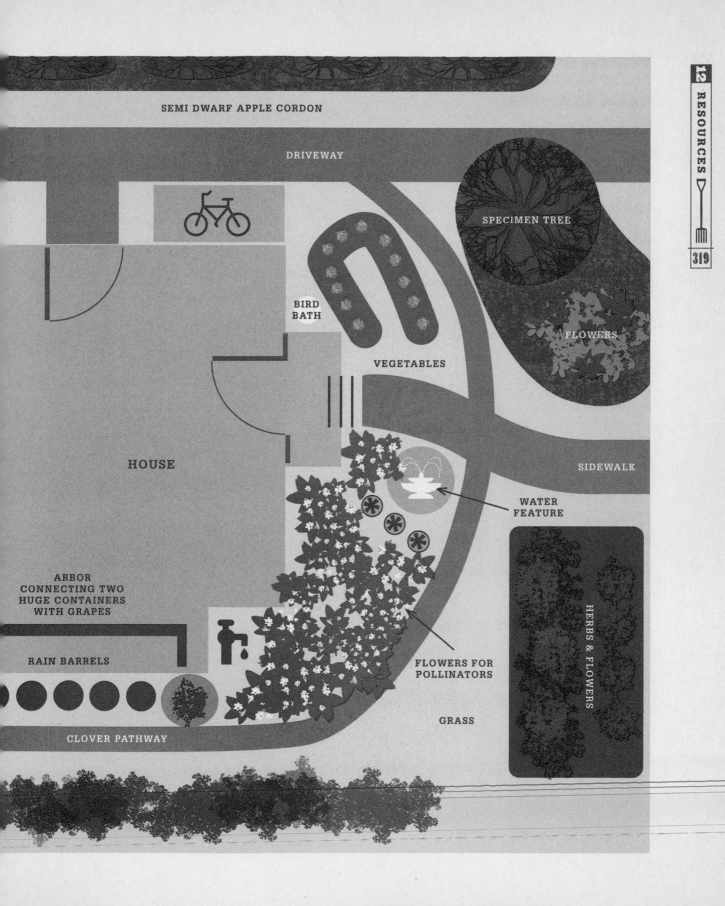

SEMI DWARF APPLE CORDON

DRIVEWAY

SPECIMEN TREE

BIRD
BATH

FLOWERS

VEGETABLES

HOUSE

SIDEWALK

WATER
FEATURE

ARBOR
CONNECTING TWO
HUGE CONTAINERS
WITH GRAPES

HERBS & FLOWERS

RAIN BARRELS

FLOWERS FOR
POLLINATORS

GRASS

CLOVER PATHWAY

Organic Gardening Glossary

Aerobic: An environment containing oxygen.

Aggregate: A group of soil particles that hold together, these are the building blocks of soil structure.

Anaerobic: An environment without oxygen.

Beneficial insect: An insect that benefits plants by eating or parasitizing pest insects or by pollinating flowers

Bilateral symmetry: The left and right sides of an organism are a mirror image of each other.

Biodiversity: A healthy environment known for the presence of myriad and diverse living and dead organisms.

Blanch: To briefly submerge fruits or vegetables in boiling water, then cool and drain.

Biomass: The total weight of plant material growing in a given area.

Bolt: The process of a plant going prematurely to seed.

Cane fruit: Fruit from plants that belong to the **Rubus** genus, includes raspberry, blackberry and loganberry.

Castings: Red wiggler or earthworm poop, also called vermicompost.

Cloche: A light-permeable cover that protects plants from frost and traps heat.

Compaction: The pressing together and pressing down of soil particles by foot or other traffic.

Compost: To compost is to use any of several methods to speed up the decomposition of raw organic matter, usually by aerating and moistening piles of materials containing carbon and nitrogen. Compost, the result of these efforts, is well-rotted organic matter. It is dark, crumbly, nutrient-rich natural fertilizer.

Cotyledon: First leaves of a plant, sometimes called seed leaves. They look immature and chubby.

Cover crop: A crop that improves and protects the soil in which it is grown.

Cultural techniques: Gardening techniques to care for and tend plants.

Dampening-off: A fungal disease of young seedlings, causing the stem to rot and the plant to keel over and die.

Deciduous: Plants that shed their leaves at the end of the growing season.

Decomposer: Soil bacteria and other larger organisms that are nourished by breaking down the remains or waste of other organisms into simple organic compounds or compost.

Determinate tomato: A short, bushy plant that grows to a determined size, sets flowers and fruits over a short period, and tends to ripen all at once.

Espalier: A small-space pruning and training technique used to grow fruit trees (or ornamentals) in a flat plane.

Evaporation: The loss of water from the surface of the soil.

Exoskeleton: A hard external shell encasing the bodies of all arthropods.

Fertilizer: Any material added to the soil to provide essential nutrients to plants.

Floating row cover: Sometimes called "remay," a spun polyester fabric that lets in sun, air and water; it is draped over plants to protect them from pests and frost.

Frass: Debris or excrement produced by insects and other arthropods.

Frost dates (first and last): The first and last day an area experiences freezing temperatures.

Gray water filtration system: To filter household waste water from the sink, shower or washing machine so that it can be used to irrigate landscape plants.

Green manure: A cover crop used to protect the soil, hold or build nutrients, and smother weeds.

Green shoulders: The top or stem end of a vegetable is green rather than the usual color – this can indicate too much sun exposure or that the vegetable is not yet ripe.

Harden-off: To slowly introduce a seedling or transplant that has been growing indoors or with the protection of a cloche or other shelter into the outdoor garden.

Herbaceous: Plant matter that is soft and green; plants that do not form woody tissue.

Hoof rot: Inflammation on the feet of sheep, goats or cattle that is caused by bacterial infection.

Hoop house: A miniature greenhouse constructed over garden beds out of flexible pipe or heavy wire and covered with clear plastic to protect plants from frost and to trap heat.

Hot water bath-canning: A method for preserving high acid foods in boiling water.

Humus: The result of organic matter binding with minerals in the soil to create a moist, deep-brown, rich, healthy tilth.

Hummus: A bean puree made from chickpeas.

Hybrid: A plant that is a genetic dead-end, meaning the seeds from a hybrid plant will not give rise to the exact same or a very similar plant.

Indeterminate tomato: Vines continue to grow, making flowers and fruit throughout the season; tall vigorous plants.

Inoculant: Spores of a desired strain of rhizobium bacteria applied in powder form to the appropriate legume before planting. Also, any material of high microbial content added to soil or compost to stimulate biological activity.

Instar: A stage of life in insects and other arthropods between exoskeleton molts.

Larva(e): The immature, wingless stage of an insect that will undergo complete metamorphosis.

Leaching: The movement (usually loss) of dissolved nutrients as water percolates through the soil.

Legume: A member of the plant family

Leguminosae: Plants (including clover, alfalfa, beans, and peas) whose roots host nitrogen-fixing bacteria in a symbiotic relationship.

Ley Crop: A crop set aside to improve the soil for two or more years; frequently cut to make compost.

Macronutrient: A plant nutrient needed in substantial quantities, including carbon, hydrogen, silica, oxygen, nitrogen, phosphorus, sulfur, calcium, magnesium, and potassium.

Mandibles: The mouth parts of insects and arthropods.

Metabolism: The biochemical processes of growth, maintenance, and energy transformations necessary for a living organism.

Micronutrient: A plant nutrient needed in very small quantities, including copper, chlorine, zinc, iron, manganese, boron, and molybdenum.

Mineralization: The release of water-soluble mineral and simple organic compounds through the decomposition of organic matter.

Nitrogen fixation: The conversion of gaseous nitrogen into complex chemical compounds that can eventually be used by a plant.

Nursery bed: A protected area in the garden where seedlings are nurtured.

Nymph: The immature stage of an insect that undergoes incomplete metamorphosis.

Open-pollinated: Unlike hybrids, plants that will return in the same form from their seed.

Organic matter: The living part of the soil comprising decomposing plant and animal materials, and huge amounts of microorganisms.

Overwinter: The process that occurs when a plant is either planted out or sown in the late summer to grow through the fall and winter months for harvest in spring.

Parasitoid: An insect which uses another insect as a nursery for its eggs.

Pomme fruit: The fruit of apple, quince or pears having edible flesh that surround several seed chambers.

Pressure-canning: A method for preserving low-acid foods that require higher temperatures than boiling water for sterilization.

Pricking out: The process of gently transplanting small seedlings from a flat into larger pots.

Pulses: Legume seeds for human consumption: beans, peas, favas, lentils, etc.

Pupa: An insect is the transformational stage between larva and adult.

Pupate: To become a pupa or to emerge as an adult insect.

Rhizobia: A group of bacteria that penetrates the roots of legumes, extracts carbohydrates from the plant, and fixes gaseous nitrogen in the soil which plants can use.

Radula: The rasping tongue of snails, slugs and other mollusks used to scrape or eat food.

Rootstock: A root that is used for plant propagation or on to which another plant is grafted.

Ruminants: Cloven hoofed, cud-chewing mammals such as cattle, goats, deer and bison.

Sheet mulch: A composting and soil building technique that smothers weeds with layers of wet cardboard or newspaper that are covered by mulch.

Sidedress: To apply granular fertilizer on the soil surface around the plant to stimulate growth.

Soaker hose: Porous hose used for irrigation that allows water to seep along the length of hose.

Soil: An ecological system consisting of space for air and water, inorganic minerals, organic matter, and living organisms.

Soil amendment: Any material added to the soil to promote biological activity.

Soil pH: The pH of the water in the soil, which controls the availability of phosphorous and trace elements, and the diversity of soil organisms. The pH for most soils ranges from 5.0 to 9.0; 7.0 is neutral and ideal for growing vegetables.

Soil structure: The density and size of soil aggregates. A good soil structure is composed of aggregates of widely varying size.

Subtropical or tender plants: Plants that cannot survive even a light amount of frost and require adequate heat to grow well.

Symbiotic: The mutually beneficial interrelationship between two organisms, such as nitrogen-fixing rhizobia and the roots of legume plants.

Tilth: An Old English word used to describe the structure and quality of cultivated soil. Tilth is similar to health. In the medieval monastic tradition, the word was used to describe the cultivation of wisdom and the spirit.

Topdressing: Compost or fertilizer applied to a growing crop, usually spread on top of the soil and only minimally mixed in.

Transpiration: Water loss through leaves.

Transplant shock: Stress or damage to plant as a result of digging it up and planting it in a new location.

True leaves: The leaves of a plant that grow after the cotyledons.

Undersow: To sow seeds underneath an existing growing crop with minimal cultivation of the soil.

Variety: A cultivated, named form of a plant species demonstrating unique characteristics.

Vegan: A vegetarian whose diet contains only plants and plant products.

Sources for Charts and Lists

MNGG = **The Maritime Northwest Garden Guide**. Seattle: Seattle Tilth, 2008.

COG = **Comprehensive Organic Gardener Manual**. Seattle: Seattle Tilth, 2009.

ACKNOWLEDGMENTS

Y HEARTFELT THANKS GO TO FAMILY and friends whose support and words of encouragement were invaluable during this process. Special thanks to my partner Joan Goodnight who graciously took on the first-reader task, gently helping me sculpt and refine gigantic ideas into simple, straightforward prose – your love and support are indispensible. Thanks to my son Alwyn for reminding me to make time for fun and play. Many thanks to my mother and step-dad, Jean and Steve Ludeman, for all the phone calls, words of wisdom and unending support. To Sherry Edwards who read early drafts and makes me laugh. I am also indebted to Nance, Mary, Aaron and the citizens of Bruvoldia for picking up the various pieces, bringing pizza and keeping the technology working.

I would also like to thank Andrea Dwyer, Executive Director at Seattle Tilth for the opportunity to synthesize more than 15 years of learning and teaching about growing food – and share it with you. I am especially grateful for Marty Wingate's contributions to the vegetable and fruit lists, urban farm profiles and for unraveling the planting calendar and first frost dates dilemma. Many thanks to Liza Burke and Falaah Jones for their encouragement, attention to detail and indomitable spirits. I appreciate the extraordinary efforts of Jessica Heiman and everyone who kept the learning gardens at Seattle Tilth growing while I was immersed in this project. My gratitude to the wonderful gardeners, educators and staff of Seattle Tilth – you are an inspiration – and to all the children and families who have taught me so much about worms, plants and life. Thanks to Dinah Dunn and her amazing team at Black Dog and Leventhal who took my words and brought them to life on these pages. Finally –

Thank You Garden!

INDEX